Naturalists' Handbooks 34

Leaf beetles

DAVE HUBBLE

T0306967

Pelagic Publishing
www.pelagicpublishing.com

Published by Pelagic Publishing
www.pelagicpublishing.com
PO Box 725, Exeter, EX1 9QU, UK

Leaf beetles
Naturalists' Handbooks 34

Series Editor
William D.J. Kirk

ISBN 978-1-78427-150-3 (Pbk)
ISBN 978-1-78427-158-9 (ePub)
ISBN 978-1-78427-159-6 (Mobi)
ISBN 978-1-78427-160-2 (PDF)

Text © Pelagic Publishing 2017

Dave Hubble asserts his moral right to be identified as the author of this work.

All rights reserved. Apart from short excerpts for use in research or for reviews, no part of this document may be printed or reproduced, stored in a retrieval system, or transmitted in any form or by any means, electronic, mechanical, photocopying, recording, now known or hereafter invented or otherwise without prior permission from the publisher.

British Library Cataloguing in Publication Data
A catalogue record for this book is available from the British Library

Cover photographs
Top: Scarlet lily beetle *Lilioceris lilii* on damaged leaf (Henrik Larsson)
Bottom left: *Donacia* sp. on yellow flag iris (Dave Hubble)
Bottom-right: Alder leaf beetle *Agelastica alni* on garden plant (Dave Hubble)

Printed and bound in India by Replika Press Pvt. Ltd.

Contents

Erratum
Fig. 4.7
p. 39

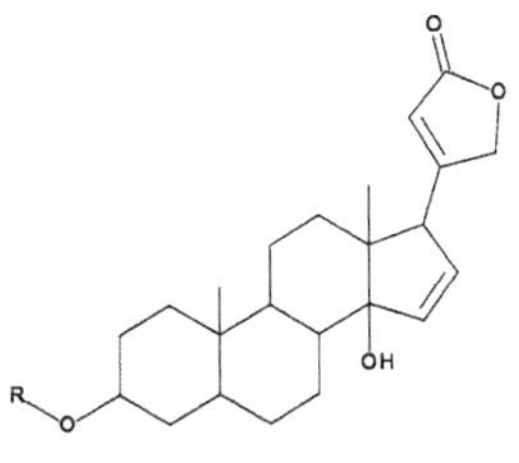

Editor's preface

Leaf beetles are a fascinating and distinctive group of plant-feeding beetles with a rich diversity of interesting features. They are not as well known as ladybird beetles, but they can match them for attractive appearance and remarkable behaviours. Many of the adults are brightly coloured or are a metallic blue, green, purple, brass or bronze colour. They include garden pests such as the lily beetle, crop pests such as the cabbage stem flea beetle and invasive pests such as the Colorado potato beetle. Some common leaf beetles are even beneficial pollinators. In contrast, there are also many rarities such as the tansy beetle and the hazel pot beetle, and species that may now be extinct in the British Isles, such as the Pashford pot beetle. There are species that exude noxious fluid or feign death to defend themselves, species that make audible squeaks and species that can jump further than fleas in relation to their body length.

This *Naturalists' Handbook* provides an introduction to the natural history, behaviour and ecology of the leaf beetles, and advice on how to find and study them, together with an identification key and a checklist.

I hope that this book will encourage more people to study leaf beetles. There is much to discover about even the commonest species and they are easy to find. A wide range of habitats support leaf beetles, including gardens, crops and urban brownfield sites. Species records can make useful contributions to mapping schemes, which can then be used to show the effects of climate change or changes in land use.

This *Naturalists' Handbook* on leaf beetles complements other titles in the series on beetles: *Ladybirds* (no. 10), *Common ground beetles* (no. 8) and *Weevils* (no. 16). Leaf beetles are also described within their habitats in *Insects on cabbages and oilseed rape* (no. 18) and *Insects on dock plants* (no. 26).

William D.J. Kirk
February 2017

Acknowledgements

First of all, my thanks go to the staff of the Biological Records Centre for all their help developing and inspiring my work as an entomologist. In particular, Helen Roy and Björn Beckmann have been instrumental in providing encouragement, access to academic reprints and other material, and giving me the initial nudge to take on the Chrysomelidae Recording Scheme.

I am equally indebted to Rebecca Farley for her generous permission to re-use material from my AIDGAP publication – essential for producing the key in this book.

Thanks are also due to Darren Mann and Amoret Spooner for access to the Oxford University Museum of Natural History's biological collection, without which this book would be much less well illustrated.

Lastly, there are more entomologists than I can possibly name whose records have helped inform this work. Their efforts are much appreciated as without them, there would be no knowledge of species distribution or many other aspects of chrysomelid life.

About the author

Dave Hubble works in Environmental Science for The Open University, where he particularly enjoys getting students away from their computers and out into the field. He has been a fan of invertebrates and their diversity since first encountering vast arrays of pinned specimens as a child visiting London's Natural History Museum. As well as various academic papers, he has published an identification key to leaf beetles in the AIDGAP series (2012) and wrote the latest Status Review of the group (2014). As well as being involved in various Working Groups covering topics such as Pollinators, he is a Fellow of the Royal Entomological Society, and organises the national Chrysomelid Recording Scheme. When not tinkering with small Coleoptera, he is a professional poet and artist, and is involved with various social and environmental campaigns.

About Naturalists' Handbooks

Naturalists' Handbooks encourage and enable those interested in natural history to undertake field study, make accurate identifications and make original contributions to research.

A typical reader may be studying natural history at sixth-form or undergraduate level, carrying out species/habitat surveys as an ecological consultant, undertaking academic research or just developing a deeper understanding of natural history.

1 Introduction

Leaf beetles are not as well known as ladybirds, but they too have many species that are distinctively patterned and dome-shaped. In addition, there are ten times more species and many of them are important crop pests. They are an attractive, fascinating and important group of beetles that deserves to be better known. The aim of this book is to introduce the leaf beetles and provide information to allow anyone to find and study them.

1.1 What are leaf beetles?

The leaf beetles are a large group with nearly 300 species in the British Isles (Table 1.1), and an estimated 40,000 worldwide in over 2,500 genera, making them one of the most diverse and regularly encountered beetle groups. As their common name suggests, they are all herbivorous and closely associated with plants, where they feed on various parts, not only leaves. Being beetles, they have biting mouthparts (unlike the piercing-sucking mouthparts of the Hemiptera or 'true bugs') and forewings that are hardened to form elytra (wing cases) covering the abdomen when viewed from above.

Most are in the family Chrysomelidae, with a small number of species in the closely related Megalopodidae and Orsodacnidae – collectively these three families constitute the leaf beetles and they are also generally referred to as 'chrysomelids', because previously, these three families were

Table 1.1 Numbers of British leaf beetles by family and subfamily

Family	Subfamily	Species in subfamily	Species in family
Chrysomelidae	Amblycerinae	1	281
	Bruchinae	16	
	Cassidinae	14	
	Chrysomelinae	45	
	Criocerinae	9	
	Cryptocephalinae	25	
	Donaciinae	21	
	Eumolpinae	1	
	Galerucinae	148	
	Lamprosomatinae	1	
Megalopodidae	Zeugophorinae	3	3
Orsodacnidae	Orsodacninae	2	2
		Total	**286**

classified together in the Chrysomelidae. It is important to become familiar with the main features of the various subfamilies as this makes identification quicker, and a brief introduction to them follows.

Amblycerinae. These are similar in appearance to the Bruchinae and also associated with legumes. They were previously grouped with them as a separate family.

Bruchinae. Previously considered as part of a separate family, bruchines have been variously known as seed beetles, pea weevils, bean weevils, bean beetles and beanseed beetles because of their association with the seeds of leguminous plants. Many are pests of such crops, including dried and stored produce, especially in tropical and subtropical areas.

tribe
a taxonomic rank below family or subfamily level but above genus

Cassidinae. Previously considered to be a tribe (Cassidini) of the subfamily Hispinae (which is not found in Britain), cassidines are now given subfamily status. They are commonly known as the 'tortoise beetles' owing to their dorsally flat-domed and more-or-less rounded appearance.

Chrysomelinae. Chrysomelines are the 'typical' (i.e. domed, relatively large and often metallic in colour) leaf beetles and include many of the more charismatic species within the British chrysomelid fauna, such as the rare tansy beetle *Chrysolina graminis*.

Criocerinae. In Britain, a small subfamily of relatively elongate, parallel-sided beetles represented by eight species including the introduced lily beetle *Lilioceris lilii*. All have a notch on the inner edge of the eye, although this may be slight.

Cryptocephalinae. A subfamily consisting of two tribes in Britain: the Clytrini (composed of one scarce and one recently extinct species) and the Cryptocephalini (all belonging to the genus *Cryptocephalus*). The genus *Cryptocephalus* includes several rare or endangered species and they are known colloquially as 'pot beetles' because of the appearance of the cocoons their larvae live in. These are initially built by the female during and immediately after egg laying, with the egg being held between the rear tarsi (feet) and covered by her faeces. Once covered, the pots are dropped to the ground among leaf litter, which often forms much of the larval diet, and the larvae add their own faecal material to their cocoons as they grow. The precise structure of these cocoons varies by species.

Donaciinae. Commonly known as the 'reed beetles' and associated with plants in or near water bodies and wetlands. Donaciines are more elongate than many other chrysomelids and many are distinctive, being brightly metallic in colour.

Eumolpinae. A small subfamily known by a single species in Britain, *Bromius obscurus*. They are superficially similar to the Chrysomelinae, but can be distinguished by features of the legs.

Galerucinae. A large subfamily comprising two tribes, the Galerucini and Alticini, both of which have previously been considered separate subfamilies. The Alticini are known as 'flea beetles' because of the well-developed flea-like jumping abilities of the adults and have been known as both the Halticinae and Alticinae. Examples include the cabbage-stem flea beetle, *Psylliodes chrysocephala*.

Lamprosomatinae. A small subfamily known by a single species in Britain, *Oomorphus concolor*. Like the Cryptocephalinae, they form cocoons from faecal matter.

Orsodacninae. In Britain, this is the only subfamily within the family Orsodacnidae. There is a single genus *Orsodacne* in Britain, consisting of two species of relatively elongate beetles. The subfamily was previously placed within the Chrysomelidae.

Zeugophorinae. In Britain, this is the only subfamily within the family Megalopodidae. There is a single genus *Zeugophora* in Britain, consisting of three species of relatively elongate beetles. It was previously placed within the Criocerinae and later the Orsodacninae, which is now a separate family, the Orsodacnidae. Some authors (e.g. Zaitsev & Medvedev, 2009*) place it (along with the Orsodacninae) within the Chrysomelidae.

* References cited in the text appear in full under authors' names in References and further reading on p. 134

1.2 Adult external morphology

Chrysomelids are small to medium-sized beetles – in Britain ranging from a little over 1 mm in length (*Mniophila muscorum* and *Longitarsus minusculus*) to 18 mm in large specimens of *Timarcha tenebricosa*. Their overall shape varies from being somewhat elongate (e.g. Donaciinae) to oval and domed (in many subfamilies) or rounded and flattened (Cassidinae) (Figs. 1.1–1.5).

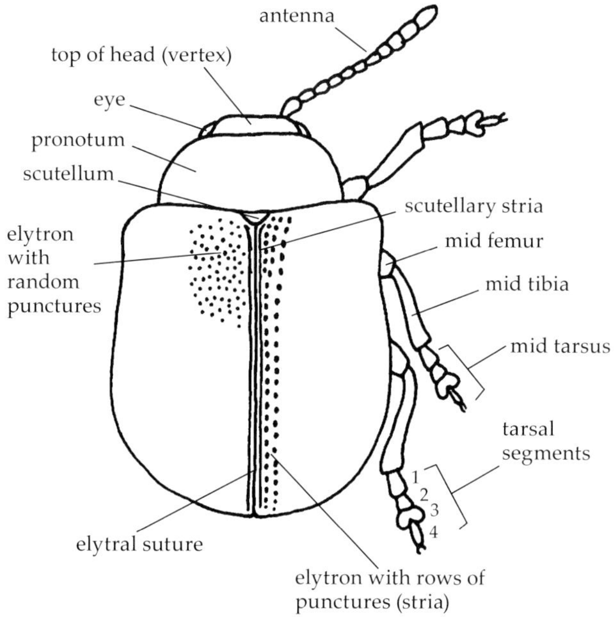

Fig. 1.1 Dorsal view of a typical chrysomelid.

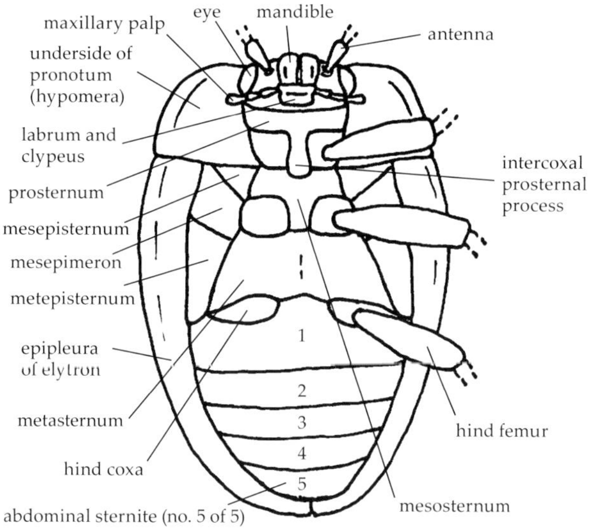

Fig. 1.2 Ventral view of a typical chrysomelid.

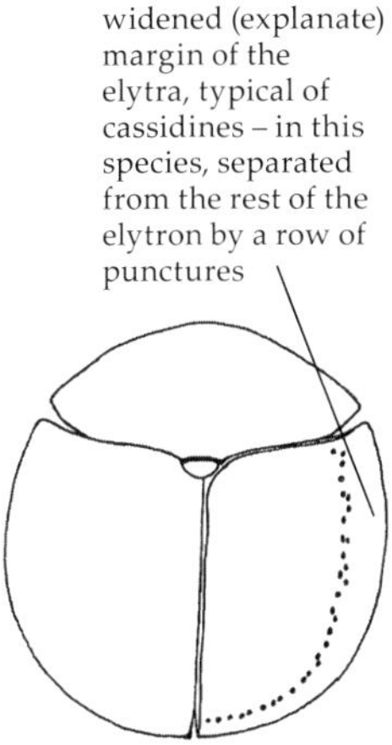

Fig. 1.3 Ventral view of a typical cassidine.

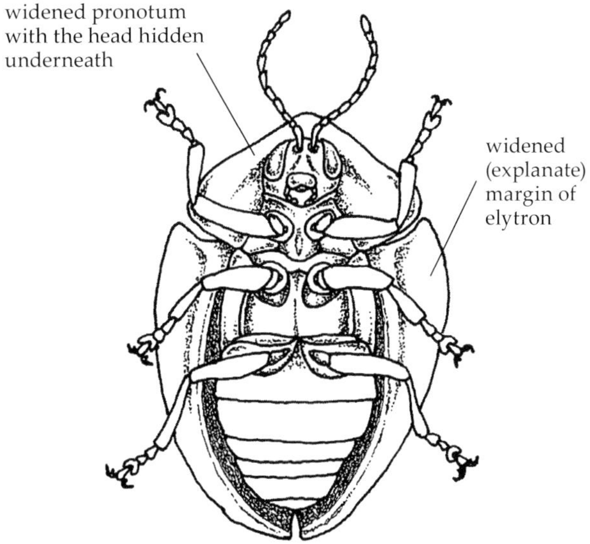

Fig. 1.4 Dorsal view of a typical cassidine.

The antennae are usually thread-like (filiform), although in some taxa (e.g. *Bruchidius*) they may be saw-toothed (serrate) or have modified and expanded segments (e.g. males of *Phyllotreta exclamationis* and *Phyllotreta nodicornis*) (Fig. 1.6). There are 11 antennal segments except in *Psylliodes* species, which have 10. The antennae are never clubbed, unlike those of the ladybirds, which some leaf beetles resemble.

Fig. 1.5 A pair of *Donacia* showing their elongate form.

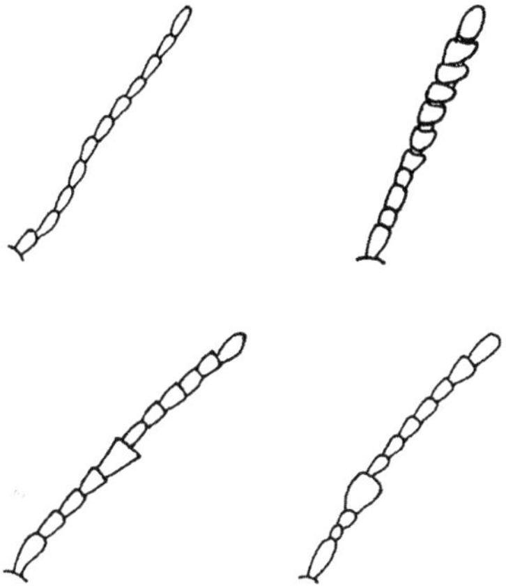

Fig. 1.6 A selection of antennal forms in the Chrysomelidae.

chitinous
made of chitin, a polymer found in the exoskeletons of many arthropods

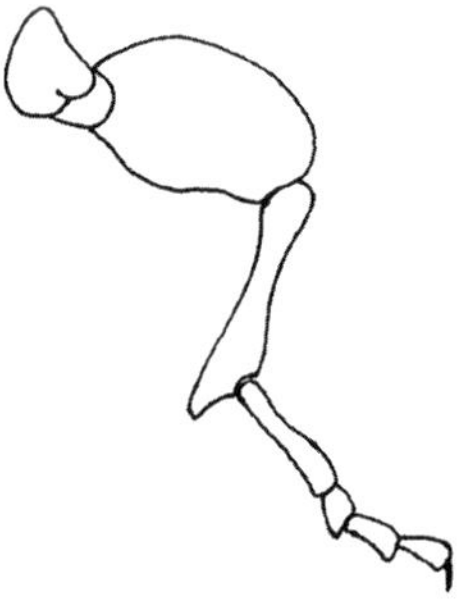

Fig. 1.7 Hind leg of a typical alticine.

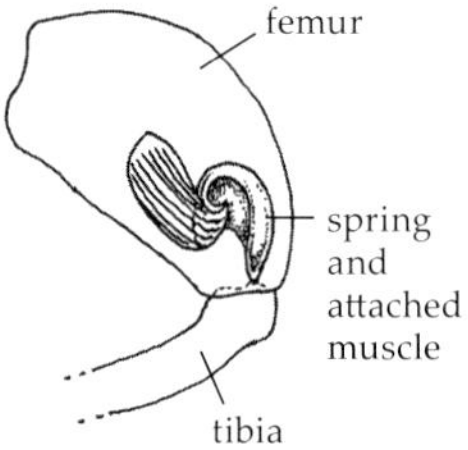

Fig. 1.8 Metafemoral spring.

In the flea beetles (Alticini), the hind femora are enlarged (Fig. 1.7) and contain a chitinous 'metafemoral spring' which allows them to jump by releasing stored energy generated by the tibial extensor muscle (Fig. 1.8). The shape of the spring varies between genera and therefore could, in principle, be useful in identification. The tarsi are all 4-segmented and in most species at least some are bilobed (if not, then the final tarsal segment is elongate as in *Macroplea*).

The head of a chrysomelid has a pair of compound eyes and mouthparts adapted for plant-feeding. The point of attachment of the antennae relative to the eyes can be useful in identification, as can the grooves and bulges of the head (Figs. 1.9, 1.10).

In most chrysomelids, the elytra cover the full length of the abdomen, although in the Bruchinae, the final segment (pygidium) remains exposed, and in some Galerucinae they may be shortened. Gravid (egg-bearing) females of some species, such as the common green dock beetle *Gastrophysa viridula*, have a swollen abdomen, which forces the elytra apart. The elytra may also be patterned as in a number of *Cryptocephalus* species. As well as aiding identification, these spotted patterns are likely to provide protection through mimicry of unpalatable species such as ladybirds (Chapter 4).

Although chrysomelids are ground and plant-dwellers rather than aerial specialists, the hindwings are used for

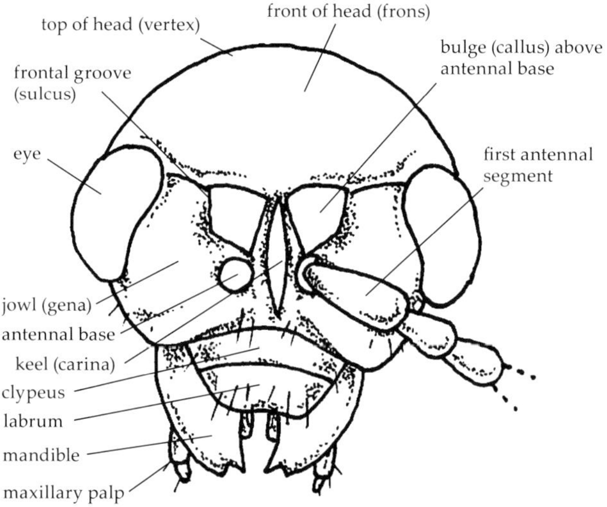

Fig. 1.9 Front view of the head of a typical chrysomelid, *Altica lythri*.

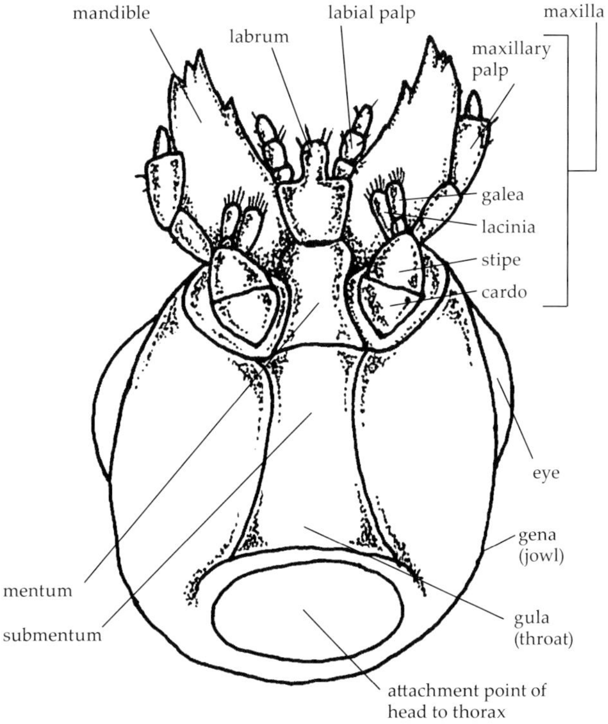

Fig. 1.10 Ventral view of the head of a typical chrysomelid, *Altica ythri*.

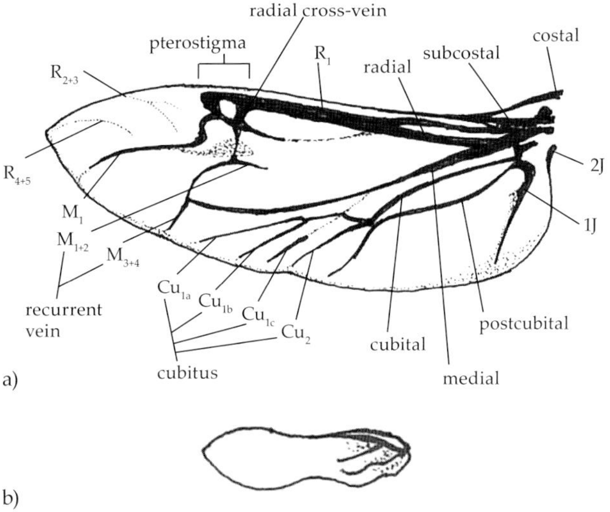

Fig. 1.11 (a) Hindwing of a fully winged chrysomelid showing venation; (b) Hindwing of a flightless chrysomelid with reduced wing development.

flight in fully winged species with developed flight muscles and associated thoracic structures (Fig. 1.11). See Chapter 3 for more about variation in wing development and flight ability.

See Chapter 2 for details of juvenile morphology, and for morphology of reproductive structures.

1.3 Evolution and palaeontology

The earliest fossil chrysomelids date from the Jurassic (200–145 million years ago) and showed a great increase in diversity by the end of the Cretaceous (approximately 65 mya) at the same time as diversification of flowering plants. The Donaciinae were the largest group until the Pleistocene (approximately 1.8 mya), although the modern genera and species became established during the Palaeogene (65–23 mya) and Neogene (23–2.6 mya). The apparent dominance of donaciines may be due to their aquatic and riparian (water edge) habitats providing anoxic sediments favourable for fossilisation, although such habitats also remained stable for millions of years. This may explain why *Plateumaris nitida* (a North American species) is almost identical to *Donacia primaeva* from 30 mya, having changed little during this period (Elias & Kuzmina, 2008).

The more recent Quaternary (2.6 mya–present) fossil record has provided several hundred chrysomelid taxa (with some but not all recognised at species level) around the world, indicating its importance as a plant-feeding beetle group. Western Europe (especially Great Britain) is relatively well studied and around 150 fossil chrysomelid species are known from this area. The majority are associated with wetland habitats, although species from deciduous woodland, meadows and grasslands are also well represented. However, not all chrysomelid groups are equally represented, with only nine species of flea beetles known from fossil resin worldwide, most of which come from eastern Europe (Bukejs, 2014).

During the Pleistocene glaciations, it is likely that dry steppe and steppe-tundra formed the dominant landscape type in open, unglaciated parts of Europe. The fluctuating conditions associated with glacial cycles, including the bare soil left by retreating ice sheets would have favoured ruderal ('weedy') plants such as thistles (*Cirsium*) and plantains (*Plantago*), which are effective colonisers. These are also plants readily fed upon by a wide range of chrysomelids and thus the terrestrial beetle species may have been favoured by these conditions.

Despite the large numbers of fossils, these still only provide incomplete snapshots of the prehistoric fauna and the origins and spread of many of the Chrysomelidae are not well understood except in fairly broad terms. However, as the evolution of plant groups provides opportunities for new taxa of plant-feeding insects to arise, it is possible to follow this process to some extent. The first evolutionary branching of the Chrysomelidae probably occurred in the Upper Jurassic (161–145 mya) as cycad-feeders expanded to exploit monocots (early criocerines) and water-lilies (donaciines) as these evolved. Modern donaciines are still associated with wetland habitats and plants, while criocerines still feed on monocots e.g. *Lilioceris* (including the lily beetle *Lilioceris lilii*) on Liliacaeae and *Crioceris* (including the asparagus beetle *Crioceris asparagi*) on Asparagaceae, having specialised and diversified alongside these plant families. Along with this shift from generalist to specialist plant-feeding, it is possible that chrysomelids developed from pollen-feeders to being external and internal feeders on a range of plant parts.

monocots or monocotyledons
plants with a single seed-leaf (cotyledon), for example grasses, orchids and lilies

Feeding relationships can therefore shed light on chrysomelid evolution, for example by looking at those that feed on crucifers (Nielsen, 1988). As with all chrysomelids, such beetles have had to adapt to the physical and chemical characteristics of their host plants, and although larvae and adults of any given species usually feed on the same plant, they tend to use different parts, with adults mainly on leaves (sometimes buds or flowers) and larvae often on or in (as miners) roots, stems and leaves. Crucifer-feeding has evolved separately several times with specialists developing early on in the tribe Phaedonini, and more recently in the Chrysomelini (both tribes within the subfamily Chrysomelinae). Plant–beetle relationships may also be used in evolutionary studies. For example, Poinar & Jolivet (2004) used information from host-plant biology and distribution and the fossil record to conclude that the ancestral food plant of *Timarcha* was a shrub in the family Ericaceae, because *Vaccinium* (an ericaceous genus including bilberry and cranberry for example) is the only host genus still used by both Old and New World beetles in this genus.

crucifer
a plant in the cabbage-family Brassicaceae (formerly Cruciferae)

Considering the present-day fauna, there is no clear distinction between the Palaearctic and Oriental/Asian regions as many genera are shared and also found in tropical Africa. The long evolutionary history has led to significant spread and some British genera may have originated in what is now Australasia e.g. the genus *Phaedon* which

Palaearctic
an ecozone (a large area where species have a long shared evolutionary history) covering Europe, Asia north of the Himalayas, north Africa and the northern and central parts of the Arabian Peninsula

shows similarities with *Geomela* (Australia) and *Aphilon* (New Zealand) (Reid, 1995; Daccordi, 1996). A number of strict relationships between chrysomelid groups and their host plants are known, such as that between *Aphthona* and spurges (*Euphorbia*), and the many specialist species-level relationships. A particularly well-defined example of the latter is the highly localised British endemic, bronze Lundy cabbage flea beetle *Psylliodes luridipennis*, which is found only on Lundy Island off the coast of north Devon, and feeds solely on Lundy cabbage *Coincya wrightii*, which is also endemic.

endemic
a species that is restricted to a particular place

2 Life history

2.1 General life history

The life cycles of leaf beetles go through the same stages as all other beetles – egg, larva, pupa, adult – and undergo full metamorphosis, a situation known as being holometabolous. In many British species, the life cycle takes one year, but this is highly variable within the leaf beetles, and even within a single species, and the following types of life cycle are known:

- One generation per year e.g. *Oulema melanopus*, *Chrysolina banksi* and *Phyllotreta undulata* (along with many other species, including most 'flea beetles') showing a peak of adults in spring or summer depending on emergence time, with one of the life cycle stages overwintering. Some, such as the lily beetle *Lilioceris lilii*, have one generation per year in Britain but up to three in warmer areas of continental Europe. Some species can be variable, such as tansy leaf beetle *Chrysolina graminis*, which shows one generation per year but around 5% of the population surveyed near York undergo a long diapause of over a year, greatly lengthening the life cycle of those individuals (Cox, 2007).

diapause
a delay in development resulting in a period of dormancy

- One generation taking more than one year. For example, *Donacia dentata* has a two-year life cycle while *Donacia semicuprea* has a three-year life cycle. Although not all are well understood, several species of *Cryptocephalus* (e.g. *Cryptocephalus coryli*) have life cycles lasting either two or three years.

aestivation
a dormant period during the summer, similar to hibernation in winter

- Aestivation e.g. the rosemary beetle *Chrysolina americana*. Adults can be found all year with a new generation emerging during May and June, then aestivating until July to September when they feed again and then overwinter. Breeding can occur from autumn through to April in southern England.

- Overlapping generations e.g. the bloody-nosed beetle *Timarcha tenebricosa*. Various generations of eggs and larvae may exist at the same time, along with adults throughout the year as they can live for 14 months or more.

The number of larval instars (stages) also varies within the chrysomelids, from three in *Timarcha* and *Phyllotreta* (among others) to six in *Cryptocephalus nitidulus* and *C. coryli*.

Most species mate frequently though some such as *Plagiodera versicolora* can mate later in the year and store sperm in order to fertilise eggs the next spring. In many chrysomelids, the life cycle is not fully understood, with the length of various stages unknown, manner of overwintering undetermined, or juvenile stages undescribed. In others, details are known from captivity but not in the wild. These are areas where useful study could be undertaken.

2.2 Eggs

Fig. 2.1 Eggs of *Gastrophysa viridula*.

Most chrysomelid eggs are elongate and oval (Fig. 2.1), ranging from approximately 0.15 mm wide and 0.3 mm long in small Alticini such as *Phyllotreta undulata* to around 1.9 × 3.5 mm in *T. tenebricosa*, the largest British species. Some are cylindrical with rounded ends (e.g. *Chrysomela tremula*), ellipsoidal (e.g. *Sphaeroderma rubidum*), highly elongate (e.g. *Orsodacne cerasi*), hemispherical (e.g. *Phyllobrotica quadrimaculata*) or globular to approximately spherical (e.g. *Pyrrhalta viburni*). Most are pale whitish to creamy or yellow, but the range of colours includes lime green, brown, orange-red, purplish, grey and black, with some more-or-less colourless and translucent. Some species show considerable variation in egg colour, which changes as they mature. For example, the eggs of *Chrysolina staphylaea* are a deep red-brown when freshly laid, becoming considerably paler, while those of *Crioceris asparagi* begin greenish-grey, darkening to become black. The eggs of most *Cryptocephalus* species are a typical yellowish colour, but this is obscured by their covering of dark faecal matter. The egg surface also varies from smooth to being extensively microsculptured or rough, with patterns and depth differing between species. The eggs of around 70 British species remain undescribed, something that careful observation of adults during oviposition could help to remedy.

oviposition
egg-laying

petiole
leaf stalk

Oviposition occurs on host plants, usually on leaves, sometimes petioles or stems. Some wetland species lay eggs on or between submerged (e.g. *Donacia marginata*) or floating (e.g. *Donacia dentata*) leaves, or under water in gelatinous cases (e.g. *Macroplea appendiculata*).

In some species, adults prepare the egg-laying surface (substrate), such as *Oulema obscura*, which lays eggs singly or in pairs on the lower surface of grass leaves, usually in grooves made by adult feeding. It is, however, unclear how predictable this behaviour is as eggs may also be laid at the bases of leaf stalks, and are not always laid parallel to the leaf veins and therefore the grooves. Eggs are attached to the substrate using a sticky, gelatinous substance, which may be visible when investigating eggs *in situ* e.g. in *L. lilii* where it is yellow.

The period of egg-laying varies greatly between species. The majority have a single brood with eggs laid over a period of a few months in spring, summer or autumn. Some have multiple broods such as *Longitarsus luridus* which lays eggs mainly in March to April and September to October. Similarly, although eggs of *Longitarsus pratensis* have been recorded from January to October, there are two peaks of adult emergence, suggesting spring and late summer oviposition, with possible overwintering of eggs.

Aside from the possibility of overwintering, eggs generally incubate for a relatively short period, from several days to a few weeks as the embryo develops. For example, Cox (1981) recorded eggs of *Orsodacne cerasi* taking between 11 and 14 days to hatch at 25–30°C. Also, females laid 20–35 eggs over a 9-day period but only had 12–15 ovarioles in each ovary, indicating that more than one egg may develop in each ovariole. In contrast, Owen (1999) noted that *Cryptocephalus coryli* laid up to 15 eggs per day over a 10- to 20-day period and that they took around 5 weeks to hatch. This is a larger species and hatching occurs within the pot that subsequently encases the larva (Owen, 1999). The number of eggs laid per female also varies between species, both in terms of the number produced per batch and the total over a female's lifetime, with some larger beetles such as *Timarcha* and *Chrysomela* surviving for two or more years.

ovarioles
tubes forming egg-production units

2.3 Larvae

Chrysomelid larvae have three thoracic segments, each with a pair of legs, and each leg terminates in a tarsal claw. The abdomen has ten segments, the last being narrowed (and sometimes extended) to form an anal proleg. In most species the overall shape is that of a 'typical' beetle larva i.e. a curved cylinder widest around the abdomen (Fig. 2.2). Exceptions include the Cryptocephalinae where the larvae are more strongly curved to fit inside their case (Fig.

Fig. 2.2 Larva of *Gastrophysa viridula*, a typical chrysomelid.

2.3), and the Cassidinae, which are flattened with spiny projections, including a caudal fork (Chapter 4). In some cases e.g. *Donacia* (Fig. 2.4), the legs are reduced to small clawed buds, which, along with its pale colour, give the larva the appearance of a 'typical' maggot or grub. Donaciines also have a pair of mobile curved spines at the tip of the abdomen (arrowed in Fig. 2.4). These are often believed to be used to pierce submerged plant parts, allowing the larvae to access air as they are fully aquatic, feeding and developing under water. However, as donaciine larvae have been seen to develop without piercing plant tissues, it is possible that the spines are simply for attachment and that, as there are no gills, dissolved oxygen is taken in through the cuticle. As the relationship between donaciine larvae and oxygen availability will depend on water quality, investigation into the mechanism of uptake may yield valuable information relevant to factors such as pollution, eutrophication and conservation management.

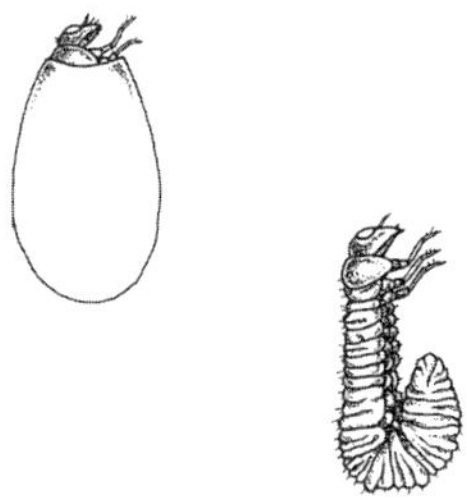

Fig. 2.3 Larva of a typical cryptocephaline.

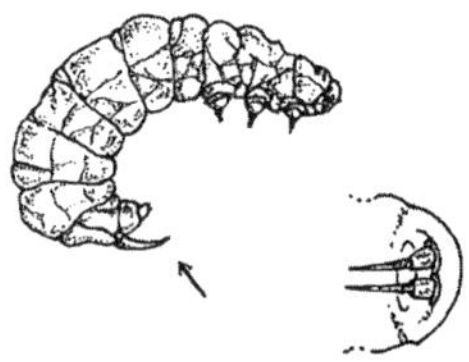

Fig. 2.4 Larva of a typical donaciine.

The head capsule (Fig. 2.5) is hardened and generally darker than other parts of the larva. The mandibles are short, as are the antennae and maxillary palps, and there are a number of ocelli on either side of the head near the antennal bases. This number varies between species and may be useful in larval identification.

ocelli
small eyes with a single lens

Many chrysomelids have 'hatching spines' or 'egg-bursters' (Cox, 1988). These are small hardened teeth or spines that grow from the cuticle of embryos or first-instar larvae (Fig. 2.6). They may be found on the thorax and/or abdomen and are used to slit the chorion or shell to aid

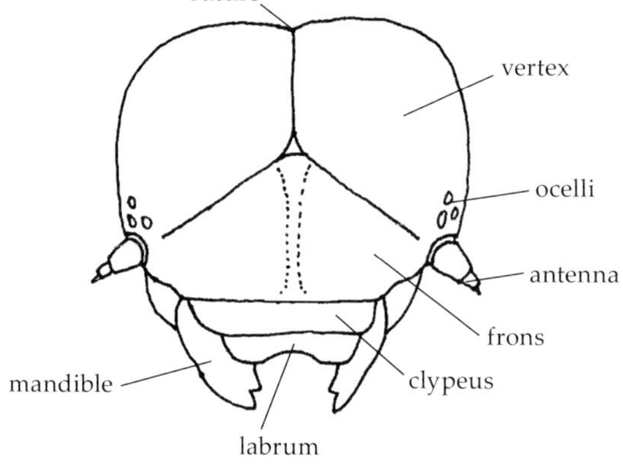

Fig. 2.5 Larval head capsule.

endopterygote insects
members of the Endopterygota, the most diverse of the insect superorders, characterised by the development of wings inside the body and an elaborate metamorphosis involving a pupal stage

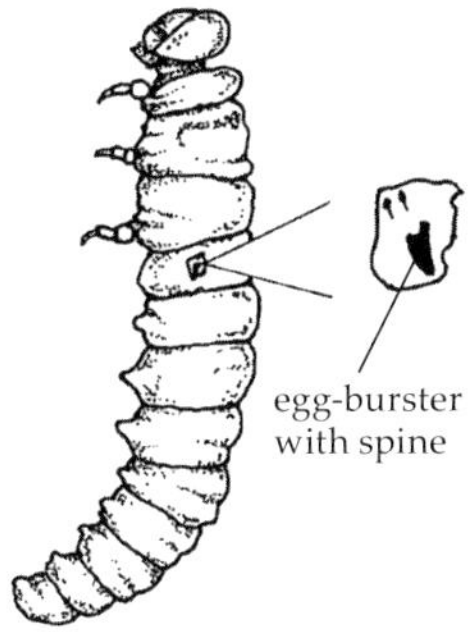

Fig. 2.6 Larval egg-bursters.

emergence. This is not the only method of hatching, even in species with egg-bursters, as chewing, pushing with the anal proleg, pushing with the head, using secretions from the mouth (which may soften the chorion around the exit hole, or lubricate the area), and whole-body movements are also employed.

2.4 Pupae

As in all beetles and other endopterygote insects, chrysomelids undergo complete metamorphosis with the mature larva pupating and emerging as an adult. The pupa generally resembles the adult more than the larva as features such as the elytra, fully developed legs, and long antennae and maxillary palps are visible (Fig. 2.7). Eight sternites and nine tergites are visible, the ninth bearing the vestiges of

sternite
underside abdominal plate (sclerite) of a segment

tergite
upperside abdominal plate (sclerite) of a segment

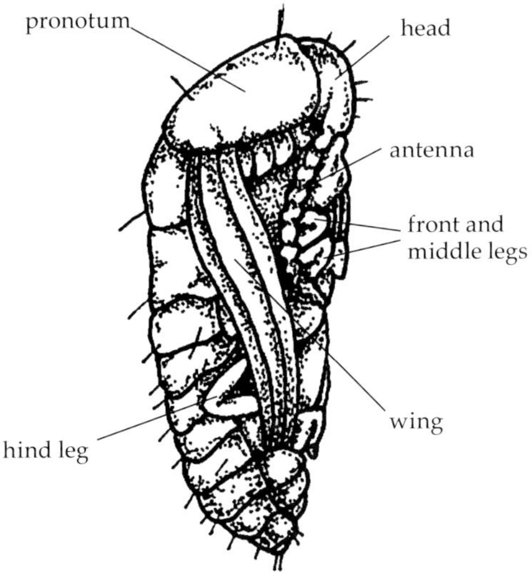

Fig. 2.7 Pupa of a typical chrysomelid.

urogomphi
in beetle larvae, a pair of outgrowths of the dorsal surface (tergum) of the ninth segment. In chrysomelids, they appear as short spines. Also known as pseudocerci or corniculi.

two urogomphi. Donaciines pupate in submerged cocoons constructed by the maturing larva, and attached to plant structures such as roots and rhizomes (Fig. 2.8). Cassidines, as with their larvae, are distinctive, being flattened with spiny outgrowths (Fig. 2.9, Chapter 4).

Although there are comprehensive works covering the identification of adults, and to a lesser extent larvae, there is little published material on identification of chrysomelid pupae. Many are known to develop in an earthen cell but are otherwise undescribed. Due to their location in the soil and

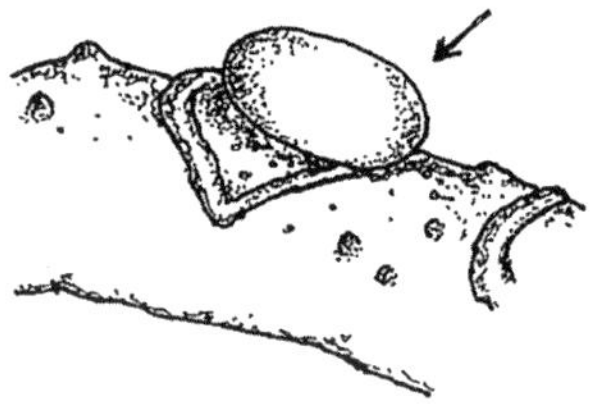

Fig. 2.8 Pupa of a donaciine.

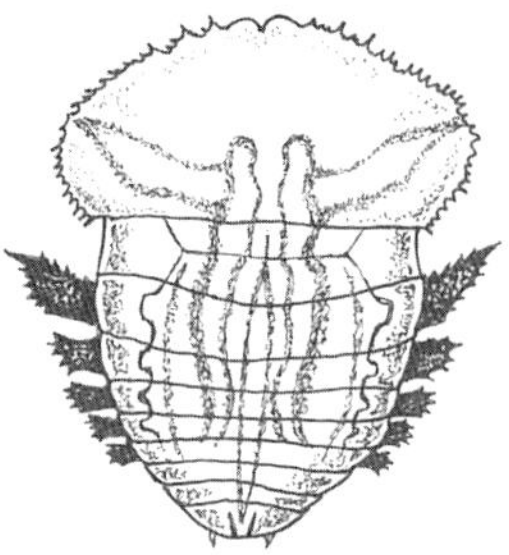

Fig. 2.9 Pupa of a cassidine.

a camouflaged cocoon, these pupae are rarely encountered without deliberate searching, hence this is an area where considerable knowledge could be gained with diligent investigation.

2.5 Adults and reproduction

The aedeagus, broadly analogous to the mammalian penis, is the male reproductive structure most often used in species identification, in particular its median lobe (a term sometimes used interchangeably with 'aedeagus'). The median lobe is a chitinous tubular structure used to transfer sperm to the female. It is associated with a partial covering called the tegmen, and an elongate filament known as the flagellum may be visible near, or at, the tip of the median lobe. The flagellum develops from the soft endophallus or internal sac within the aedeagus, and its function may be as a 'guiding rod', to prepare the female's genital opening for copulation, or it may be directly involved in sperm transfer, because in some species such as *Lilioceris lilii* the flagellum appears to be tubular with an apical opening (Düngelhoef & Schmitt, 2006). Parameres (lateral lobes) of the aedeagus are absent from most chrysomelids, but are present in the Megalopodinae, Orsodacnidae, Bruchinae, Donaciinae, Sagrinae (absent from Britain) and *Timarcha*. Within these taxa the parameres can be single or paired, and vary in both length and number and density of setae. Hubweber & Schmitt (2006) discovered tiny holes in the parameres of *Donacia* and suggested that they may be glandular openings, unlike the sensilla on the median lobe which each have a minute seta and are therefore likely to have a sensory function. Whether sensory or secretory, the functions of these setae and openings are not yet known and are therefore areas for further study. A schematic diagram of the male genitalia is given in Fig. 2.10.

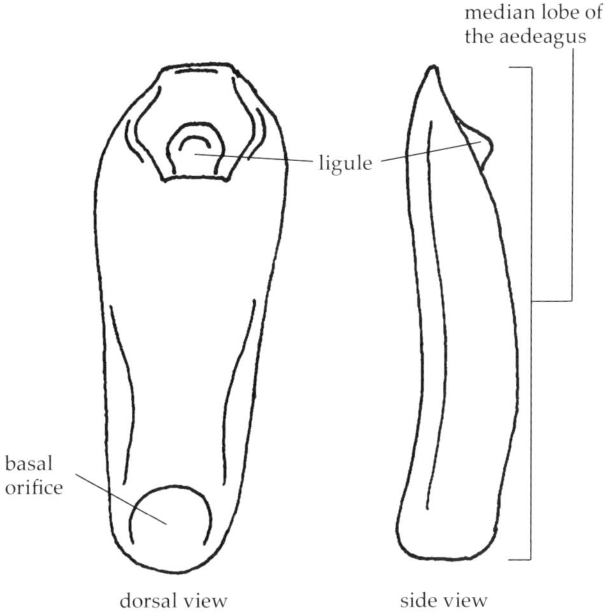

Fig. 2.10 Median lobe (or aedeagus) of male chrysomelid genitalia.

Fig. 2.11 Female *Gastrophysa viridula* showing the swollen abdomen associated with egg-bearing in some species.

sclerotised
hardened by the formation of a protein called sclerotin and often darker in colour

In females (Fig. 2.11), the reproductive structures may fill much of the abdominal cavity and consist of a pair of ovaries, each leading to an oviduct, and these join to form a common oviduct, which opens into the genital chamber. There is also a spermathecal organ consisting of a capsule, duct and accessory gland, which also opens into the genital chamber. In some subfamilies, the ovipositor also has a pair of accessory glands which secrete a viscous fluid onto the eggs for adhesion or protection. Female reproductive structures are used less often for identification than those of males as the lack of sclerotised structures makes it difficult to find clear and consistent differences between taxa. However, at least in some cases the shape and relative length of the spermatheca and spermathecal duct are useful as diagnostic features because their structure is known to be species-specific. A schematic diagram of the female genitalia is given in Fig. 2.12.

Overall, although the morphology of reproductive structures is generally well understood, their functions during copulation are not, and are likely to vary between species. A greater understanding of such functions can inform areas such as sexual selection and conflict. For example, females of the neotropical cassidine *Chelymorpha*

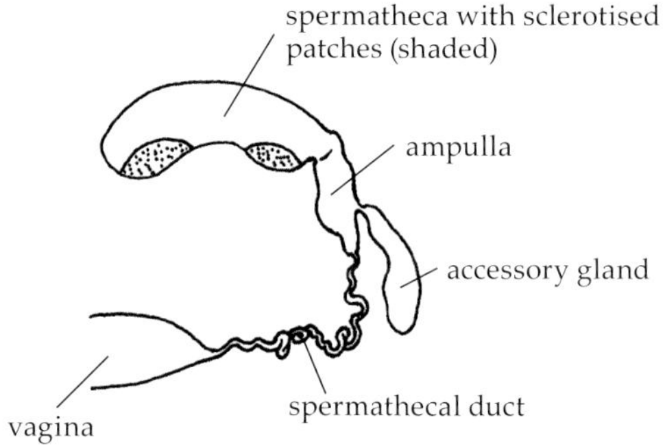

Fig. 2.12 Female chrysomelid genitalia.

alternans prefer males with a longer flagellum, while male *Callosobruchus maculatus* pierce the wall of the bursa copulatrix (part of the female genital tract) to prevent her from mating again. Criocerines, however, achieve a similar end by mate-guarding (Düngelhoef & Schmitt, 2006). These examples challenge the assertion that most chrysomelids mate frequently.

While most chrysomelids lay eggs, some are partly or wholly ovo-viviparous. For example, *Gonioctena decemnotata, Gonioctena pallida* and *Gonioctena viminalis* are ovo-viviparous in Britain, though in southern Europe they lay eggs. In *Gonioctena olivacea*, the strategy is less certain but it appears to retain eggs so that they hatch almost immediately once laid and so may be ovo-viviparous. There may also be variation within a genus as *Chrysolina varians* is ovo-viviparous in Britain while others of the genus are not. Viviparity occurs more often at higher latitudes and altitudes as a mechanism to reduce the time required for external larval development in locations where sufficiently warm summer periods are short.

ovo-viviparous
producing eggs inside the body, but giving birth to live young, which hatch prior to laying

3 Leaf beetles in their environment

Leaf beetles have the potential to be excellent indicators of environmental conditions because they have a close relationship with their hostplants and most species feed on a restricted host range. For example, links between aquatic pollution and declining populations of donaciines are well documented (e.g. Menzies & Cox, 1996). The close beetle–plant relationships also mean that, while some chrysomelids may be pests, or prove invasive if they are imported with foodplants, others may be useful in biological control of weeds.

3.1 Habitats and foodplants

All leaf beetles feed on plants and their mouthparts are adapted for this with mandibles for chewing. Although mouthparts are rarely used in the identification of adults e.g. the neotropical work of Cabrara & Durante (2003), larval identification is more likely to use features of the mandibles to separate species.

In most species, larvae and adults feed on the same plants, though there are exceptions such as *Luperus longicornis* which feeds on grass roots as a larva but moves to the leaves of trees and shrubs when adult. Being more mobile, adult leaf beetles can feed on a wider range of plant parts, though most favour leaves, while larvae tend to be more restricted, with many being root feeders, apart from chrysomelines and cassidines which tend to feed on leaves. Larvae of a few species feed on dead material, such as *Cryptocephalus parvulus* which prefers old, mouldy birch leaves, while myrmecophilous beetles such as *Clytra quadripunctata* utilise decaying material in ants' nests, and *Smaragdina affinis* may also do so, though this is uncertain. A small number of species do not feed on vascular plants. For example *Labidostomis tridentata*, now considered extinct in Britain, feeds on algae including that found growing on tree bark, and *Mniophila muscorum* probably feeds on the mosses where it is usually found. In Britain, the only species with fruit-feeding larvae is *Lochmaea crataegi*, which mines the pith of hawthorn fruit. There are several seed-feeding larvae, all within the Bruchinae. A number of these are accidental introductions with bulk foods and almost all feed on members of the pea family (Fabaceae), and thus

myrmecophily
a relationship with ants that is commensal (where one organism benefits without affecting the other) or mutualistic (mutually beneficial)

Fig. 3.1 *Hydrothassa marginella* feeding on pollen (flower centre).

can damage stores of commodities such as lentils and chickpeas. Adult bruchines, along with those of several other genera feed at least partly on pollen, though further observations, including examination of gut contents, are required to confirm this in many species and to determine the plant species involved. Leaf beetles are commonly seen on flowers (Fig. 3.1), but this does not necessarily indicate pollen-feeding as they may feed on nectar or parts of the flower itself. This also highlights their role as pollinators, which has not been thoroughly researched, although several *Phyllotreta* species are known to be important pollinators of both wild and cultivated members of the cabbage family (Brassicaceae), as well as feeding on them.

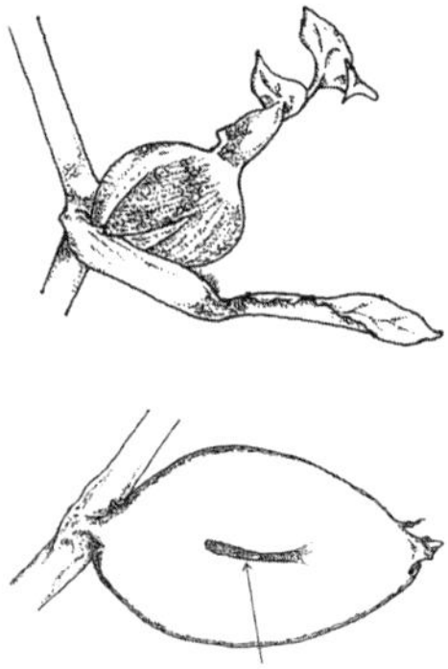

Fig. 3.2 Gall caused by a chrysomelid.

No British species cause galls, but some do elsewhere such as the Asian *Oulema reclusa* which causes galls at the tips of growing stems of the dayflower *Commelina paludosa* (Vencl & Nishida, 2008) (Fig. 3.2). This is unusual not only because few chrysomelids are gall-causers, but also because the host is a monocot unlike the great majority of gall hosts. Living within a gall, the larva is morphologically much reduced with little pigmentation or sclerotisation, no larval shield or associated setae as usually found in criocerines, no ocelli and neither legs nor ambulatory warts. Thus it is soft, blind and poorly mobile, but is protected by the gall that also provides its food. There are, however, a number of plant miners in the British fauna (Chapter 7), including all three *Zeugophora* species, which mine poplar leaves, and many *Psylliodes* species and other flea beetles, which mine various parts of their respective foodplants, especially stems, although galleries may extend from there to other structures such as petioles and leaves. Leaf-mining in the strict sense is fairly rare in British chrysomelids. In a few

ambulatory warts
small, blunt projections used for locomotion

cases, plant mining may occur, but has not been confirmed, as in *Longitarsus strigicollis* (formerly *Longitarsus fowleri*), *Longitarsus ganglbaueri*, *Longitarsus holsaticus* and *Longitarsus tabidus*, which may mine roots. Most root-feeding larvae live and feed on the outside of roots, though some such as *Podagrica* mine mallow roots, and some flea beetles, such as *Longitarsus exoletus* and *Hermaeophaga mercurialis*, have been confirmed as root-miners, although the former produces generally superficial mines, and the latter only mines larger roots.

For many species with root-feeding larvae, although the foodplant is known, the details of larval behaviour, including feeding, have not been recorded in detail as the larvae are hidden underground. There are also a few species whose larval feeding location remains unknown such as *Orsodacne cerasi*, *Orsodacne humeralis* and even the common and widespread *Neocrepidodera transversa*.

The association between beetle and foodplant of course begins prior to hatching. Although egg-laying often occurs on parts of the foodplant, this is not always the case. For example, *Phyllotreta striolata* sometimes lays eggs in soil near, or attached to, roots. This means the eggs and newly hatched larvae are initially less conspicuous, but the protection afforded is a trade-off against the risks of having to move once hatched.

There are a number of examples where British species, some very familiar, can be used in the biological control of food plants that have been introduced outside their native range. For example, heather beetle *Lochmaea suturalis* has been introduced into New Zealand since 1992 to control heather *Calluna vulgaris*, which is a serious invasive weed in parts of the country such as Rotorua and Tongariro National Park, having been introduced historically to create artificial grouse habitat, but is now outcompeting and dominating native plants. The effectiveness has been mixed, with significant heather damage as intended in Rotorua, but little impact at Tongariro. The reasons for this difference are unknown and could be due to subtle effects of climate, selection and rearing methods creating a genetic bottleneck, a lack of nitrogen in the heather leaves, or other factors.

Similarly, the green dock beetle *Gastrophysa viridula* has been evaluated for biological control of docks, which can be weeds, either when invasive elsewhere, or in their native range if grasslands are poorly managed (e.g. Martinková & Honěk, 2004). However, although the beetle can skeletonise leaves and reduce plants' vigour sufficiently to prevent them setting seed, docks are usually not killed and other measures may be needed in combination such as encouragement of the rust fungus *Uromyces rumicis* (also considered to be a biological control agent) and careful management of nitrogen inputs.

3.2 Pest species and control

The Chrysomelidae includes a considerable number of pest species although few are especially problematic in Britain. At the time of writing, only two are included on the Food and Environment Research Agency (Fera) list of notifiable pests that have actually been recorded in Britain; Colorado potato beetle *Leptinotarsa decemlineata* (Figs. 3.3, 3.4) and western corn rootworm *Diabrotica virgifera* (Fig. 3.5). Other notifiable pests not yet found in Britain are also listed, namely several other *Diabrotica* species and North American potato flea beetles of the genus *Epitrix*, especially *Epitrix tuberis*. The two notifiable species that have been found in Britain have not become established due to quarantine and eradication procedures and, at least in *L. decemlineata*, the inability of adults to overwinter and complete their life cycle. *D. virgifera*, however, is more likely to establish if it can avoid eradication as its native range has similar climatic conditions to those in Britain. There are other more minor non-notifiable pests that have been recorded but failed to establish. For example, *Xanthogaleruca luteola* is found on elms in mainland Europe and *Luperomorpha xanthodera* is a Chinese species known occasionally from garden centres where it is accidentally imported via the plant trade.

Fig. 3.3 Colorado potato beetle adult.

Fig. 3.4 Colorado potato beetle larva.

Fig. 3.5 Western corn rootworm *Diabrotica virgifera*.

Several introduced bruchids are pests of legumes such as beans, peas and lentils in warmer parts of the world, but can occasionally be found associated with dried foods in Britain. These include the pea beetle *Bruchus pisorum* (Figs. 3.6, 3.7), dried bean beetle *Acanthoscelides obtectus* and azuki beanseed beetle *Callosobruchus chinensis*. Such species are mainly found in and around storage facilities where temperatures are more favourable and most are unlikely to be able to survive long term or breed in the wild, although specimens are sometimes collected outdoors. Some bruchids of commercial importance in Britain are native, such as the bean seed beetle *Bruchus rufimanus* which is a pest of various stored beans and is a widespread, if localised, species known from a variety of habitats.

Many commercial brassicas such as cabbages and water-cress are fed on by chrysomelids (Fig. 3.8). Several *Phyllotreta* species feed on oilseed rape, as does the cabbage-

Fig. 3.6 *Bruchus pisorum.*

Fig. 3.7 *Bruchus pisorum* feeding damage.

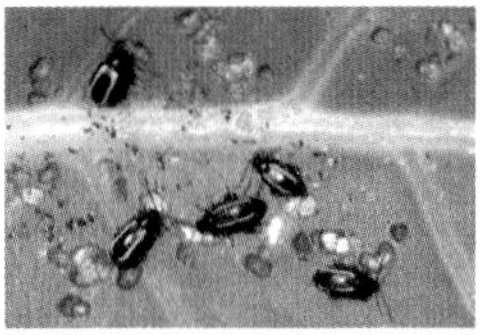

Fig. 3.8 *Phyllotreta vittula* on cabbage.

stem flea beetle *Psylliodes chrysocephala*, which can also be a significant pest of kale. As well as direct damage, feeding activity can transmit plant diseases such as turnip yellow mosaic virus and 'black rot', a bacterial disease of brassicas. In North America, cucumbers, muskmelons, squashes and pumpkins are damaged by striped cucumber beetle (*Acalymma vittatum*) and spotted cucumber beetle (*Diabrotica undecimpunctata*). These feed on the plants and are vectors of *Erwinia tracheiphila*, the bacterium causing cucumber wilt disease which spreads via feeding wounds, being carried on mouthparts or in frass (insect faeces). As the bacterium overwinters in the gut of the beetles, transmission by feeding and excretion are both effective. As a larva, *D. undecimpunctata* is known as the southern corn rootworm and feeds on various crops including corn (maize), while the adult also utilises a wide range of crops including beans and cotton as well as cucumbers. *A. vittatum* by contrast is restricted to cucumbers and related foodplants.

In Britain, flax and linseed cultivation (Fig. 3.9) increased in recent decades, peaking in the early to mid-1990s and consequently so did the range and abundance of its two main pests, *Longitarsus parvulus* and *Aphthona euphorbiae*. Since this peak, flax and linseed cultivation have declined and data on the above pest species would be useful to confirm whether they have declined in turn. The well documented spread of lily beetle *Lilioceris lilii* (Fig. 3.10) is more likely to be due to factors such as warmer conditions associated with climate change than increases in foodplants. An Asian

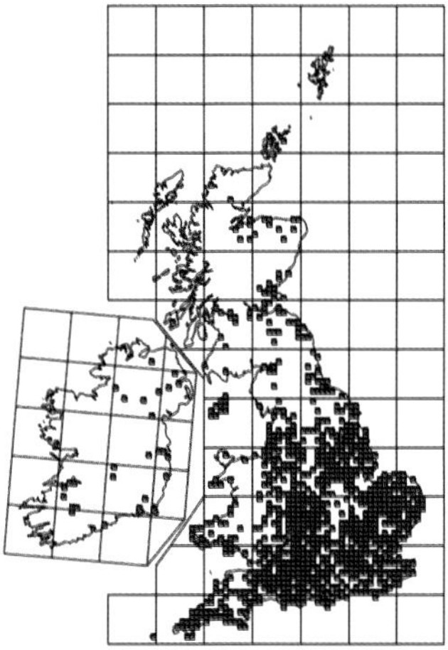

Fig. 3.9 Distribution of flax cultivation in Britain.

Fig. 3.10 Lily beetle.

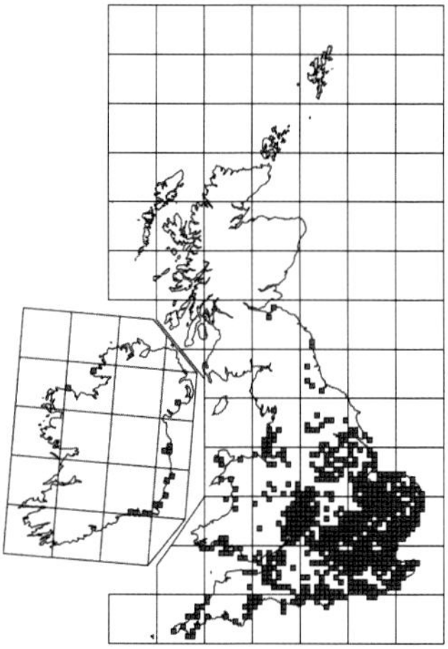

Fig. 3.11 Distribution of asparagus in Britain.

species first reported in Britain in 1839, it took a century to become established but did not start expanding rapidly until the 1990s. It is now a widespread garden and horticultural pest, though sometimes more a nuisance than a serious commercial threat, and is rarely found in wild situations despite the availability of foodplants. A member of the same subfamily, the asparagus beetle *Crioceris asparagi*, can be a significant pest of cultivated asparagus (Figs. 3.11, 3.12), and like *L. lilii*, both larvae and adults can completely defoliate plants. Similarly, *Lochmaea suturalis* can defoliate heather, including that on managed grouse moors. It may therefore be considered a pest in such habitats although, as noted in Section 3.1, its heather-feeding activity is deemed positive when used in biological control of invasive heather overseas. Some historical pests have declined along with cultivation of their associated crop, such as the hop flea beetle *Psylliodes attenuata* and henbane flea beetle *Psylliodes hyoscyami*.

Fig. 3.12 Asparagus beetle.

Some chrysomelid pests are controlled using pesticides, especially on larger commercial scales. It is beyond the scope of this work to provide a detailed account of all pesticides used to control chrysomelids, but many different chemicals have been researched and utilised with varying levels of success. Treatments may target a particular stage in the life cycle, depending on susceptibility, or may be applied to crop seeds and so control the beetle population without further application of chemicals, as for a number of flea beetles. In some cases such as *L. decemlineata* in North America, the target species has developed resistance to a range of pesticides. Most potential pests in Britain rarely reach sufficient numbers to cause serious problems, but there have been pesticide impacts on non-target species such as those feeding on arable weeds as well as some in nearby habitats (e.g. *Chrysomela tremula* in adjacent woodlands) or those in downstream water bodies, such as some donaciines.

A number of non-chemical alternatives are available and may be used in organic agriculture and horticulture. For example, companion planting of Chinese southern giant mustard (*Brassica juncea* var. *crispifolia*) has been used to control brassica-feeding flea beetles in North America by drawing the beetles away from the main crop. Physical methods such as covers, barriers and sticky traps may also be used, with yellow and white traps particularly effective. There may be potential biological control methods using predators or parasites, but more research is required into effectiveness e.g. which inter-species relationships are narrow enough to be useful. However, commercial nematode treatments are available. Of course, growing crops in a healthy and diverse ecosystem, avoiding large areas of monocultures could itself provide the companion plants, predators and parasites required to limit pest numbers.

Further information on invasive species can be obtained from the GB Non-Native Species Secretariat (NNSS); contact details are given in Chapter 8.

3.3 Variation and polymorphism

One obvious area of variation within a species is colour and pattern, especially on the elytra. Many species vary in terms of their overall colour. For example, the metallic-coloured *Chrysolina varians* can be green, blue, purple, brass, bronze or blackish, and similar colour ranges are seen in various metallic species. More sombrely coloured species also show such a range, with many flea beetles, such as *Longitarsus rubiginosus*, varying from yellowish through orange, red and dark brown. Thus colour is often a poor identification character. Patterns such as spots may be more useful, but even these can cause confusion. *Cryptocephalus bipunctatus* usually has orange-red elytra with a variable black band (Fig. 3.13), but its *thomsoni* variant has almost completely black elytra and is very similar to *Cryptocephalus biguttatus* (Fig. 3.14). Other colour variations may also be seen, such as on the head, pronotum and appendages, including the colours of individual antennal segments which can be important for identification e.g. in some bruchids.

Fig. 3.13 *Cryptocephalus bipunctatus* pattern variants.

Fig. 3.14 *Cryptocephalus bipunctatus* var. *thomsoni* and *Cryptocephalus biguttatus*.

Polymorphism in colour can be adaptive regarding thermoregulation and reduced penetration by UV radiation. When Mikhailov (2008) investigated *Oreina* and *Crosita* species, he found that darker specimens absorbed more solar radiation, reflected less and thus warmed more effectively while decreasing the damage caused by UV light. This leads to darker forms being found in cooler conditions at higher altitudes and latitudes.

Flight ability also varies greatly in leaf beetles. While many species are active fliers with fully developed hindwings and associated musculature, some such as *Mniophila muscorum* are wingless and thus unable to fly, and in *Timarcha* the elytra (forewings) are fused. However, these degrees of wing development can be seen as the opposite ends of a spectrum of wing development. *Aphthona atrocaerulea* has slightly shortened wings (i.e. slightly shorter than the abdomen) but are still likely to be able to fly. *Phyllotreta exclamationis* can be fully winged, or short-winged with wings half to three-quarters the length of the abdomen (Cox, 2007) and the latter form is flightless. Several species of *Longitarsus* are flightless, such as *Longitarsus absynthii*, which has greatly reduced, sometimes vestigial, wings and *Longitarsus anchusae*, which has vestigial wings or none at all. Some species have a variety of forms from wingless to fully winged and stages in between, for example *Longitarsus exoletus*. There can also be variation with geography and sex as in *Longitarsus ballotae*, which is usually wingless in continental Europe, but can be winged or have reduced wings in north Africa, while in Britain their wings can be absent, vestigial or fully developed. The situation is complicated further by observations in Britain that found no fully winged males, suggesting that either winged males are rare, or winged specimens in Britain are females which are more difficult to identify with certainty. Such variability is not restricted to the flea beetles as the much larger *Chrysolina banksi* also has a number of different wing development forms.

Some species are capable of flight but observations indicate they are largely inactive fliers. One example is *Agelastica alni*, although it is unclear whether flight from continental Europe has been involved in their recent expansion of range in southern England. If so, they may be more active than previously believed. Also, some species are fully winged and probably able to fly, but have not been observed doing so. This includes aquatic species such as *Macroplea mutica*, which rarely leaves water, and numerous other donaciines, which tend not to leave the vicinity of water bodies. Observations of beetles in flight, or capture in interception traps could confirm flight ability in some winged species. In others, being fully winged may not mean they are able to fly, or they may do so only weakly and rarely. This is because some, such as *Chrysolina herbacea*, do not have fully developed flight muscles or associated structures of the exoskeleton.

Flight ability affects how species are recorded. For example, use of flight interception traps is unlikely to collect non-flying species. However, such methods have caught both *Apteropeda orbiculata*, which is wingless, and *Hermaeophaga mercurialis*, which is short-winged (Cox, 2004). These flea beetles may have jumped into traps if they were placed close enough to vegetation, but it also raises the intriguing possibility that such species may have fully winged forms that have not yet been recorded. Careful checking of captured specimens may be able to confirm this. For the effect of flight ability on distribution, see Chapter 5.

The causes of variations in wing development are not fully understood but are likely to be due at least partly to habitat stability, with more stable habitats reducing the need to be able to fly to new suitable areas. However, even stable habitats may change eventually, and this may be one reason why flightless forms have some winged individuals as a form of insurance against disaster for that population. The control of wing development is complex but in *Callosobruchus maculatus* is determined after the embryonic stage by temperature and seed humidity, with cooler and drier conditions leading to development of more flightless individuals. This is likely to be related to seasonality in Africa associated with the development of its foodplant, the cowpea *Vigna unguiculata* (in Britain this species is found more often with stored chickpeas *Cicer arietinum*).

3.4 Non-native species

As noted above, there are several non-native pest species in Britain. Some, especially bruchids, are transported accidentally with imported food, while others may arrive with goods such as ornamental plants. However, many others arrive naturally, either by gradual spread from continental Europe or more quickly depending on weather conditions such as strong winds or storms. These do not necessarily cause problems, nor are they always considered invasive. For example, *Longitarsus obliteratoides* is a south-west Palaearctic species first recorded in Britain in the 1960s.

Feeding on thyme, it remains scarce in a few coastal sites in Wales and Cornwall, though in continental Europe it is found inland, including on mountains. The distribution implies it is restricted by temperature with Britain being one of its more northerly locations.

In some cases, it is less clear whether populations have arrived here from the continent or been present but overlooked. One species where this process is currently under way and highly dynamic is the alder leaf beetle *Agelastica alni*. This was considered either extinct or possibly an occasional immigrant until recorded in the Manchester area in 2004. It has since been found at a number of locations in northern England and during 2014 was recorded at various sites in Hampshire as part of a wider, ongoing population expansion (Hubble, 2015). This suggests two separate populations and it is unclear whether the northern one was overlooked while the southern one colonised from the Continent, or if both recolonised or were overlooked. It is known from most countries in Europe, and in the UK there is plenty of habitat such as alders in open wetlands, although many of the most recent UK records are from urban situations. It is unknown whether this reflects a change in the behaviour of the species, or a change in habitat availability and observations of *A. alni* are therefore valuable in helping understand its distribution. For other species there is even less information, such as *Smaragdina salicina* (Fig. 3.15) which is found across much of Europe, but only known in Britain from a single male in 2010 (Hubble & Murray, 2011). It is possible that others have arrived, but to date searches have not discovered further specimens.

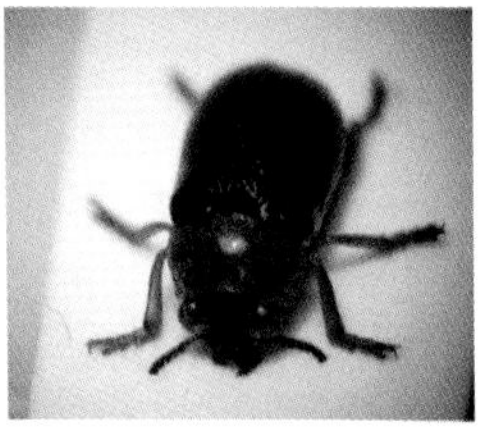

Fig. 3.15 *Smaragdina salicina.*

3.5 Conservation and threats

Legislation to protect individual species may not be effective for most species due to the practicalities of implementation and enforcement, even if sufficiently precise habitat requirements are known. Instead, conservation measures are more likely to relate to habitats and species assemblages which in turn can benefit biodiversity more broadly, beyond the targeted rarities. In a few cases such as the hazel pot beetle *Cryptocephalus coryli*, targeted survey and conservation work may be undertaken, which for this species involves searching the tree canopy in Sherwood Forest from a cherry-picker platform. The UK Biodiversity Action Plan (BAP) lists 11 chrysomelid priority species of which seven are in the genus *Cryptocephalus*. This list was last updated following the 2007 Species and Habitat Review (BRIG, 2007) and now informs the UK Post-2010 Biodiversity Framework (JNCC & Defra, 2012).

The status of species, or at least how threat and scarcity or rarity are assessed and perceived, also changes over time. The first Red Data Book to cover insects (Shirt, 1987) listed 34 chrysomelid species as either Endangered (RDB 1), Vulnerable (RDB 2), Rare (RDB 3) or Extinct. These threat levels were updated by Hyman & Parsons (1992) with 45 species covered by the previous RDB categories, plus Indeterminate (RDB I), Insufficiently Known (RDB K) and Endemic. A number of species were also listed as Nationally Scarce (Na or Nb) to give an indication of rarity within Britain rather than an assessment of threat. The most recent status review (Hubble, 2014), uses International Union for Conservation of Nature (IUCN) threat criteria, listing 45 species as either Vulnerable, Endangered, Critically Endangered, Critically Endangered (Possibly Extinct) or Regionally Extinct. A further five are Near Threatened and seven are considered Data Deficient. There is also an assessment of national rarity with two separate categories, Nationally Rare and Nationally Scarce. The Atlas (Cox, 2007) spurred an increase in recording effort and in the light of resulting data, the 2014 status review downgraded five species from Nationally Scarce (Nb) to Least Concern as they had either increased in range or were previously assessed as rarer than subsequent information suggests is the case. A further 18 species, including 10 bruchids, were categorised as Not Applicable and thus not assessed for scarcity or rarity as they are not considered native to Britain. These include some very recent arrivals such as *Bruchidius imbricornis* and *Chrysomela saliceti*, both first recorded here in 2012, plus the more familiar lily beetle which is certainly neither rare nor threatened.

Where species have declined, this is often due to loss, degradation or neglect of habitat. For example, deciduous or mixed woodland species may be affected by many factors such as felling for agriculture or development, adjacent intensive land use (such as agrochemical drift) (Fig. 3.16), conversion to conifer forestry, fragmentation, pressure from the activities of an increasing human population (Fig. 3.17), or neglect such as cessation of coppicing (Fig. 3.18) and other traditional management practices. Similar impacts may be seen in other habitats, such as pollution of fresh waters, development and neglect of heathlands, increased nutrient levels in grasslands (including meadow 'improvement') and draining of wetlands.

Fig. 3.16 Bare soil associated with intensive cultivation adjacent to semi-natural habitat.

Fig. 3.17 Damage to woodland ground by human leisure activity.

Fig. 3.18 Derelict coppice.

As well as these pressures associated with development and intensification of land use, there are also impacts associated with climate change, even if these are currently at an early stage. For example, increasing temperatures may lead to a northward spread of some species, both those already found in Britain and others which could arrive from continental Europe. Also, some potential pests such as the Colorado potato beetle *Leptinotarsa decemlineata*, which are currently unable to complete their life cycle in Britain, could become invasive if warmer conditions allowed them to become established.

However, records from beyond the existing northerly extent of a species do not necessarily indicate a true expansion in range. For example, a fairly recent record of the usually southern *Hermaeophaga mercurialis* from Cumbria could suggest a movement north except that there is already an old record from the Outer Hebrides. Therefore the Cumbrian record might simply be an individual from an overlooked population, it could be a one-off driven by weather conditions, or it might be part of a northward re-expansion following a previous reduction in range. Similarly, *Altica carinthiaca* is usually found in southern England, but as well as becoming more abundant in the south, was recorded in Cheshire in 2011. In such cases, more records are needed, and expansions can only occur when food plants and suitable habitat are also present. The lily beetle may provide a clearer example of such an expansion as it has certainly moved further north and west since the 1980s despite being present in Britain since the 19th century. As it does not seem likely that its foodplants have expanded in range, higher temperatures may well have played a part in this spread. Others include *Altica lythri* and *Phyllotreta nigripes*, both of which have spread measurably through northern England, while *Longitarsus dorsalis* and *Longitarsus flavicornis* have spread further northwards in parts of eastern England (Cox, 2007).

Temperature is not the only factor, so attributing expansions or reductions in range to climate change requires careful analysis and interpretation. For example, the flax and linseed flea beetle *Longitarsus parvulus* has expanded northwards and westwards since about 1990, but this is most likely due to wider cultivation of its food plant, allowing it to spread into new areas. However, being able to fly and feeding on other plants as well, it is likely that temperature has played some role in this expansion. If non-native species arrive here but their parasites, parasitoids or specialist predators do not, they may expand rapidly due to the competitive advantage afforded by a lack of natural enemies if other conditions are favourable.

It is unclear whether this has played a part in the spread of lily beetle, although a number of parasitic wasps do use its larvae as hosts.

Of course, climate change is more often seen as having a negative impact on species, and for good reason. Although some species may expand their range, there are likely to be wider detrimental effects on habitats. These may be because of changing weather patterns leading to droughts, floods or ongoing unpredictability, as well as the effects of sea level rise causing erosion, inundation or salinisation of coastal sites such as beaches, dunes, saltmarshes and cliffs. The spread of some species may also increase competition leading to impacts which are hard to predict.

Some species are likely to be impacted directly, such as *Phratora polaris* which is found only at altitudes between 700 and 1,100 metres on mountains in north-west Scotland. Warmer temperatures are likely to allow competitors to move to higher altitudes, pushing *P. polaris* upwards until there is nowhere left to go. It is not hard to imagine that this species might be pushed to extinction by climate change, along with others outside the Chrysomelidae that share similar habitat requirements.

4 Natural enemies of leaf beetles

Leaf beetles are particularly vulnerable to natural enemies because they are exposed on the leaf surface. Since leaves are very low in protein, leaf beetles need to spend a large amount of time feeding, digesting and largely immobile and thus prone to predation and detection by parasites and parasitoids (Pasteels *et al.*, 1988). Also, leaf damage and faeces may attract predators and parasitoids (Weselok, 1981). Anti-predation mechanisms are therefore highly developed in all stages from egg to adult. As the full life history of many chrysomelids is not fully described, there are consequently many enemies and defences that are also unknown, especially relating to juvenile stages, and these areas offer interesting opportunities for further study and research.

et al.
is short for *et alia* (and others)

4.1 Predators, parasites and parasitoids

Ants are major predators of eggs, while larvae and pupae are often targeted by spiders and true bugs (Hemiptera). Parasites and parasitoids of juvenile stages are mainly small wasps, especially in the families Eulophidae and Chalcididae. There are relatively few recorded instances of vertebrates taking chrysomelid prey, especially in Britain. However, in Poland, the common frog *Rana temporaria* (Stojanova & Mollov, 2008) and shrews (Churchfield & Rychlik, 2006) are known to do so, while the ocellated lizard *Lacerta lepida* has been recorded as a predator in Spain (Castilla *et al.*, 1991). It is likely that chrysomelids are widely taken by a range of insectivores, although direct observations are rare and detection of arthropod prey fragments is not straightforward. Adult cassidines appear to have few natural enemies, with passerine birds being the major predator (as they are large enough to remove strongly attached beetles, but small enough for such small food items to be worthwhile), and the small number of parasites being mainly tachinid flies and mermithid nematodes.

passerines
a large, diverse group of small birds including warblers, sparrows, flycatchers and tits among many others

Cox (1996) notes that there may be a succession of predators throughout the year. On broom, *Gonioctena olivacea* is predated by the mirid bug *Heterocordylus tibialis* from emergence until mid-July, then another mirid *Orthotylus virescens* until mid-September, followed by earwigs at the end of the season. However, there are more general patterns of predation overlying this with the bug *Anthocoris nemorum* known to feed on larvae of *Galerucella lineola, Agelastica alni* and *Chrysomela aenea* as well as *G. olivacea*. Chrysomelids may also be taken by species usually associated with other prey. For example, although the 2-spot ladybird *Adalia bipunctata* generally feeds on aphids, it is also known to take eggs and larvae of several chrysomelids.

Predators may mimic prey either aggressively or defensively. The blue shieldbug *Zicrona caerulea* feeds on the larvae of several *Altica* species and aggressively mimics the adults, allowing it to approach without triggering defence mechanisms. An example of defensive (and aggressive) mimicry is the way in which species of the carabid beetle genus *Lebistina* mimic and predate the highly poisonous flea beetles in the genus *Diamphidia*. The *Lebistina* larvae attach themselves to mature *Diamphidia* larvae, staying attached when they make their cocoons, ready for pupation. Then, inside the cocoon, they feed on the haemolymph and soft tissues of the *Diamphidia* larvae, killing them, and sequestering their toxins so they also become poisonous, gaining the flea beetle's defence mechanism.

4.2 Diseases, microorganisms and fungi

endoparasitic living parasitically within the host

Although many endoparasitic fungal pathogens are known to affect insects, a smaller number than might be expected appears to affect chrysomelids. However, *Beauveria bassiana* (the asexual form of *Cordyceps bassiana*) affects the widest range of chrysomelid species (Fig. 4.1), with germinated spores penetrating the cuticle, growing internally and killing the insect, emerging as a white mould-like growth with numerous small round clusters of more-or-less globular or drop-shaped asexual spores (conidia). Affecting a smaller range of species, *Metarhizium anisopliae* acts similarly but has a yellow-green spore mass, while several other fungal pathogens such as *Lecanicillium lecanii* affect a much smaller number of chrysomelid hosts (Humber, 1996). Interestingly, the major chrysomelid fungal pathogens are common and ubiquitous in insects worldwide – none are restricted, even partially, to the Chrysomelidae. However, these pathogens are highly likely to prove to be species complexes and it is therefore possible that if resolved into separate species, some may be found to be associated more strictly with chrysomelids.

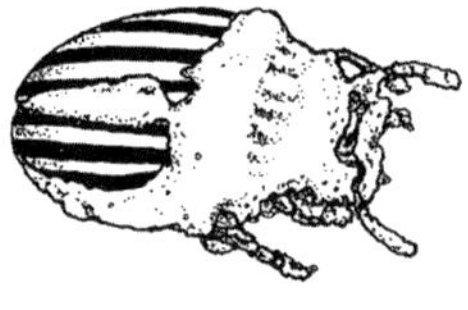

Fig. 4.1 The fungal pathogen *Beauveria bassiana.*

Previously thought to be protozoans, microsporidians are specialised microscopic fungi that are parasitic within cells (Fig. 4.2). As well as being highly host-specific, they are also often associated with particular host tissues, such

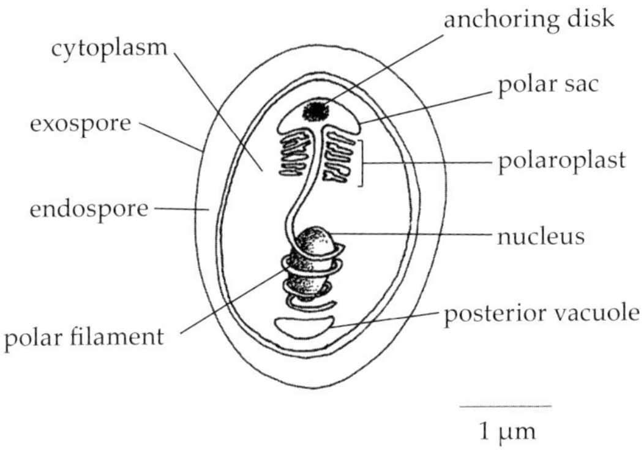

Fig. 4.2 A typical microsporidian.

polar filament
a tubular filament coiled around the nucleus of the microsporidian cell, ending in an anchoring disc

as the gut wall, fat body or reproductive system (Roy *et al.*, 2013). Once established, microsporidians can spread to adjacent cells, although the mechanism of infection is unknown. Several hypotheses have been put forward, such as 'hitch-hiking' in blood cells or using their long polar filament to bypass nearby cells and release sporoplasm (cytoplasm from a spore) in other tissues (Toguebaye *et al.*, 1988). Effects on hosts include localised tissue damage, enlargement (hypertrophy) of infected cells, reduced egg fertility due to the death of embryos, irregular growth rate (as parasitised larvae may stop feeding), and in most cases, the death of the host. Defence mechanisms against microsporidians are unknown as are the effects on chrysomelid populations. Few UK examples are known, although *Nosema phyllotretae* has been recorded from the fat bodies of *Phyllotreta atra* and *Phyllotreta undulata*. Although *Nosema* is the most commonly encountered genus, others include *Unikaryon*, *Pleistophora* and *Microsporidium*, with life cycles varying widely in complexity and form.

Chrysomelids also host a number of ectoparasitic (external) fungi, in particular the Laboulbeniales (Fig. 4.3), an order of fungi comprising around 1800 species in 200 genera. These are associated with a diverse range of terrestrial arthropods, and unlike the endoparasites noted above, are more (though not entirely) host-specific. Balazac (1988) and Humber (1996) list the recorded host-fungus associations, all of which involve just two fungal genera, *Laboulbenia* and *Dimeromyces*. Most of these are tropical, although *Laboulbenia temperei* has been recorded from *Chaetocnema aerosa*, *Chaetocnema arida* and *Chaetocnema hortensis* in France. They do not kill the host, but evidence from the harlequin ladybird *Harmonia axyridis* indicates that

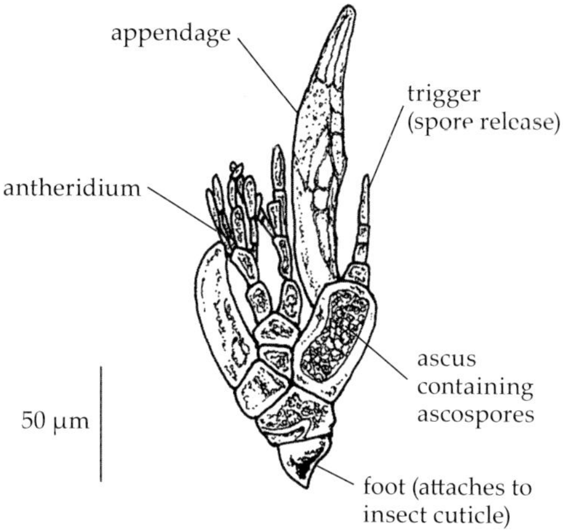

Fig. 4.3 *Laboulbenia bilobata* fungus from a South American species of *Lema*.

if infections are heavy, they may impact mating, feeding, foraging and flight (Nalepa & Weir, 2007).

gregarines
a group of large (around 0.5 mm in length) parasitic single-celled organisms

neogregarines
microsporidian fungi belonging to the order Neogregarinida

Many chrysomelid species also host gregarines (Fig. 4.4) in their digestive tracts. Again, most records are tropical or subtropical, although *Gregarina munieri* (by far the commonest) and *Gregarina crenata* have been found in nearly 100 species within several subfamilies on all continents (Théodoridès, 1988). The rate of infection can be high e.g. a study in Turkey found 38% of nearly 700 adult *Phyllotreta atra* infected with gregarines (Tosun *et al.*, 2008). Most are harmless to their hosts, although the neogregarine group is pathogenic by destroying fat tissues. However, no neo-gregarines are yet known from chrysomelids.

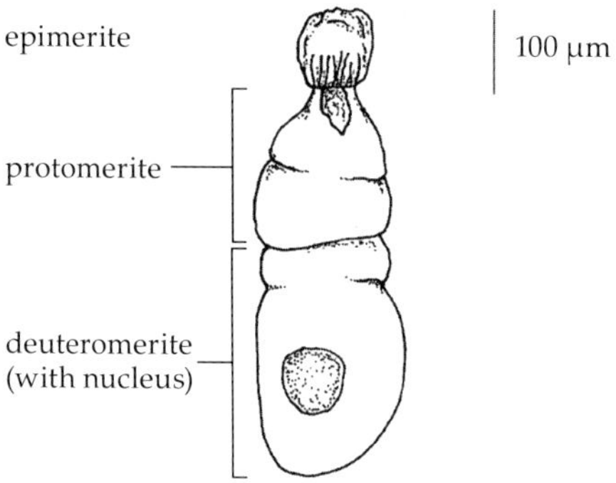

Fig. 4.4 A typical gregarine.

Nematodes are another significant parasite group, especially the family Mermithidae, but also several species of *Howardula* (family Allantonematidae) and a few from the order Rhabditida (Poinar, 1988). Infection route varies by group – larval mermithids penetrate the cuticle of beetle larvae, while in *Howardula* this is done by fertilised adult females, and rhabditids enter as larvae through existing openings such as the mouth, anus and spiracles. Nematodes may develop fully within a larva and leave before it pupates, or may be retained into the adult stage, which aids their dispersal, with some nematodes overwintering within hibernating beetles. In most cases, the host dies once the nematodes exit, although adults may live on for a few days. Up to 13,000 juvenile *Howardula* have been found in a single adult *Diabrotica*, and heavy infections may kill the host directly or render it infertile. As a defence, chrysomelids' blood cells adhere to newly penetrating nematodes, forming a capsule which hardens, immobilising and killing the parasite. In some cases, the nematodes may migrate to a part of the host where encapsulation cannot occur e.g. *Filipjevimermis leipsandra* which develops within a ganglion in *Diabrotica*; if the nematode does not reach one, it is encapsulated.

ganglion
a mass of nerve cells, analogous to the brain

Chrysomelids contain not only pathogens, but also symbiotic or mutualistic (mutually beneficial) micro-organisms. As well as providing protection (see below), maternal faeces may also transfer symbiotic (or other) micro-organisms to the larva upon hatching; certainly endosymbionts have been found in secretions applied to cassidine and donaciine eggs. The functions of such micro-organisms in the Chrysomelidae remain unknown in most cases, although in the reed beetles *Macroplea appendiculata* and *Macroplea mutica*, endosymbiotic bacteria found in the Malpighian tubules seem to be associated with secretion of a material used in the beetles' water-tight cocoons (Kölsch *et al.*, 2009).

symbiotic
showing a close and often long-term interaction between two or more species

Malpighian tubules
a system of branching tubules extending from the gut, absorbing water, waste and dissolved substances from the haemolymph and releasing them into the gut as a variety of compounds

Degrugiller (1996) also notes the presence of '*Rickettsia*-like organisms' which are symbiotic bacteria living within the cytoplasm of cells. These are generally considered to be harmless symbionts, though in *Diabrotica* they may be associated with reproductive disorders such as sperm abnormalities and reduced fertility. Similarly, mollicutes (bacteria without cell walls) have been isolated from pest species of *Diabrotica* and *Leptinotarsa* (Klein *et al.*, 2002) and although they may be gut-inhabiting commensals and

commensals
species where one benefits without affecting the other, either positively or negatively

haemolymph
the fluid in an insect circulatory system, analogous to vertebrate blood

Cucurbitaceae
a family of plants including squash, courgette, pumpkin, cucumber, watermelon and some gourds

Fig. 4.5 *Timarcha* reflex bleeding.

isoxazolinone
a class of compounds often having anti-bacterial properties

ethanolamine
a toxic, corrosive and viscous amino alcohol with an odour similar to ammonia

thus harmless to the beetles, are associated with numerous diseases of crops.

4.3 Chemical defences

One of the best-known examples of a chemical defence is the reflex bleeding that gives *Timarcha* species their common name of 'bloody-nosed beetles'. When perceiving a threat, *Timarcha* exudes a fluid consisting of haemolymph and noxious chemicals, which flows out through pores in the exoskeleton (Fig. 4.5). These chemicals include bitter cucurbitacin compounds, which accumulate in the haemolymph following feeding on Cucurbitaceae, as seen in many species including non-specialist cucurbit-feeders such as the western corn rootworm *Diabrotica virgifera* subspecies *virgifera* (Tallamy *et al.*, 2005). Although less commonly observed due to their small size, many species within the subfamily Galerucinae also show reflex bleeding as both adults and larvae, while juveniles of the African genus *Diamphidia* have a toxic haemolymph used as arrow-poison by hunters from the San people of southern Africa.

Defensive glands have been found in the larvae of a number of leaf beetle groups, including many species within the subfamily Chrysomelinae, the genus *Phratora* and some *Gonioctena*. The precise function of these glands is often poorly understood, although the location on the larval body relates to defensive behaviour – for example, in *Gonioctena*, the glands are found near the rear of the body, and the associated behaviour involves raising the tip of the abdomen. As well as acting as a chemical deterrent, there is a physical effect where the fluid clots quickly as in *Diabrotica* larvae (Wallace & Blum, 1971). This may hinder smaller predators or parasitoids, as well as potentially dislodging them when the fluid is expelled under pressure, an area where focused observation could provide useful insights. In many adult Chrysomelinae (including some common British species such as *Chrysomela populi*, *Gastrophysa viridula* and *Phratora vitellinae*), the defensive compounds are isoxazolinone derivatives, while in *Chrysolina* species, they are usually ethanolamine (Fig. 4.6) or cardenolides (Fig. 4.7) (Pasteels *et al.*, 1988). Eggs, larvae and pupae each have their own ranges of defensive chemicals which may differ from those found in other stages, or be very similar. In the genus *Chrysomela* and *Phratora vitellinae* for example, eggs contain salicin (a bitter phenolic glycoside compound found in the plant family Salicaceae, which includes willows and

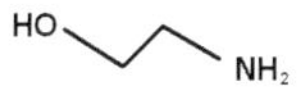

Fig. 4.6 Molecular structure of ethanolamine.

poplars) derived from hostplants by the mother, as well as isoxazolinone derivatives.

Trees in the Salicaceae vary in the concentrations of phenolic glycosides in their leaves, and some beetles show feeding preferences related to this. For example, *Phratora vulgatissima* and *Galerucella lineola* prefer leaves with low concentrations (e.g. Glynn *et al.*, 2004). In contrast, specialist herbivores such as *Chrysomela populi*, *Phratora vitellinae* and *Gonioctena decemnotata* prefer relatively high concentrations (Ikonen, 2002) as their larvae sequester the glycoside salicin and its derivatives, forming salicylaldehyde (Köpf *et al.*, 1998). This protects against generalist predators such as ants (Wallace & Blum, 1969) and spiders (Palokangas & Neuvonen, 1992), but is ineffective against specialist predators and parasitoids, such as eumenid wasps (Gathmann & Tscharntke, 1999), phorid flies (Zvereva & Rank, 2004), hoverflies (Gross *et al.*, 2004) and the sawfly *Tenthredo olivacea* (Pasteels & Gregoire, 1984), which targets leaf beetle larvae as it is attracted by their defensive secretions. Larval gut contents may also be used defensively, either via regurgitation (Fig. 4.8) or defaecation.

Fig. 4.7 Molecular structure of a cardenolide.

cardenolide
a type of steroid, many of which are toxic, and used as chemical defences by many organisms, the most familiar being the monarch butterfly (*Danaus plexippus*), which obtains the compound from its foodplant, milkweeds (*Asclepias* species)

phenolic glycoside
a chemical defence against generalist herbivores (Donaldson & Lindroth, 2007)

Alder-feeding leaf beetle larvae have interesting interactions with their host plants and their arthropod predators, such as spiders and predatory bugs. Feeding damage to alder leaves induces chemical responses which lower their quality as food, and in turn reduce the reproductive success of *Agelastica alni*, but not *Chrysomela aenea* (Baur & Rank, 1996). *A. alni* does, however, have more effective chemical defences such as reflex bleeding, although this may incur a high reproductive cost in females by slowing the larval development of their offspring. Sites with favourable temperatures can offset this cost, as can higher quality food if it can be found, given the feeding effect noted above. *Chrysomela aenea*, with weaker defences (despite the presence of glands secreting noxious chemicals, these defences are less effective), avoids predation by feeding in shadier, better-hidden areas or at higher altitudes, and by completing the vulnerable period of larval development more quickly.

Fig. 4.8 Asparagus beetle larva producing a regurgitated droplet as defence.

4.4 Behavioural defences

Like many leaf-feeding insects, chrysomelid larvae are generally actively gregarious, although the intensity of this behaviour varies between species from loose clusters in *A. alni* to tight, permanent groupings in *Phratora* (Grégoire, 1988). Apart from ovo-viviparous species such as in the genus *Gonioctena*, grouping begins with the tightly packed arrangement of eggs. As eggs tend to be brightly coloured with chemical defences, such grouping may make them a more conspicuous deterrent to larger predators, while offering 'safety-in-numbers' protection against smaller predators and parasites. Certainly, most if not all gregarious leaf beetles also have chemical defences. Pasteels *et al.* (1986) showed that eggs' defensive compounds are found in internal fluids rather than as part of the shell; thus, at least one egg needs to be pierced for the defence to take effect, but after this, the entire cluster may be deemed unpalatable and left unharmed.

It appears that newly hatched ('neonate') larvae are more likely to survive if in a group, although there has been only limited research, e.g. on the willow-feeding species *Plagiodera versicolora* by Wade & Breden (1986). As well as lacking the defensive benefits similar to those in egg clusters, isolated larvae may suffer from the 'establishment mortality' seen in other insect groups such as sawflies and moths (Ghent, 1960; Tsubaki, 1981) where individuals are unable to pierce the leaf surface to start feeding. This may be because, in groups, weaker larvae can use feeding sites initiated by stronger individuals, or because cooperative efforts are required to enable feeding. In any case, after this initial establishment, isolated larvae are able to feed and grow as effectively as those in groups. Some individuals may benefit from tight grouping through cannibalism as seen in *P. versicolorea* where feeding on sibling eggs or larvae increased growth and survival rates (Breden & Wade, 1985). Also, when resting, the larvae of some species exhibit cycloalexy, forming a tight ring with either the heads or abdominal tips pointing outwards, a position which allows defensive actions to be coordinated such as chemical secretions, regurgitation and threatening postures. In a few species, such as the scarce *Gonioctena decemnotata* (Goidanich, 1956), maternal care has been reported where the mother appears to guard her young. This species is ovo-viviparous, bearing fully formed first-instar larvae, and although the extent and mechanism of protection is unknown, the adult is conspicuously orange-red with black spots, which may be warning colouration conferring benefits to nearby young as well as the adult.

stridulation
sound production by the rubbing together of body parts

Other adult behaviours include stridulation as in the lily beetle *Lilioceris lilii* and other Criocerinae, *Clytra quadripunctata*, *Cassida viridis* and *Zeugophora flavicollis*. This is sometimes audible to humans as in *L. lilii* which produces a squeaking sound by rubbing a file-like series of small ridges on the 8th abdominal segment against small conical denticles (tooth-like projections) under the tips of the elytra. As shown by Schmitt (1988, 1994), it is more likely that stridulation in the Criocerinae is used to disturb predators and parasitoids than in communication and/or sexual behaviour, although the precise function remains unclear. Several, mainly brightly coloured, species show death-feigning (thanatosis) when disturbed, retracting their legs and falling to the ground where they lie still for some time (Fig. 4.9). One example is again the lily beetle *L. lilii*, its red colour making it conspicuous when feeding whereas it may be hidden on the ground, especially if it lands with the black underside facing up. Other responses include simply moving to the opposite side of a leaf or stem, or flying away, particularly in hot weather (Cox, 2007).

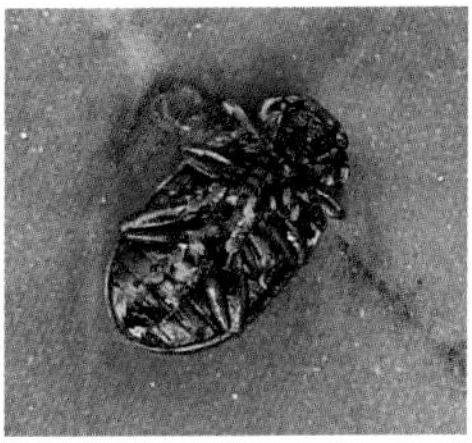

Fig. 4.9 Male *Gastrophysa viridula* feigning death.

myrmecophily
a relationship with ants that is commensal (where one organism benefits without affecting the other) or mutualistic (mutually beneficial)

A number of leaf beetle species exhibit behaviours related to their interactions with ants. Among the British fauna, the best-known example of myrmecophily is seen in *Clytra quadripunctata*, although many clytrines have similar life histories. This species lays its eggs near the nests of several ant species in the genus *Formica*, especially the wood ant *Formica rufa*, and very occasionally the yellow meadow ant *Lasius flavus*. The eggs are coated in a hardened mix of mucilage and faeces and are dropped from low vegetation near ant nests. Once the eggs are carried back to the nest by ants, which possibly mistake them for plant material due to the coating, they hatch and the larvae feed on plant debris that the ants collect. The curved abdomen is protected by a shell-like case of plant material and faeces which is added to as the larva grows; pupation takes place within this. During moults and pupation, the case is attached to twigs within the nest. Once emerged from the case, new adults leave the nest carefully, stopping when touched by an ant and exuding a drop of noxious fluid from the head by reflex bleeding. As well as avoiding predation and parasitoids, *C. quadripunctata* larvae in ant nests are protected from cold winter conditions and experience a more rapid spring temperature increase than would occur outside. The endangered *Smaragdina affinis* may also exhibit myrmecophily, but this is unconfirmed.

4.5 Structural defences

As well as the warning colours shown by species with chemical defences, mimicry is also seen in leaf beetles as an adaptation to reduce predation. A number of examples of Batesian and Müllerian mimicry relationships have

Batesian mimicry
a situation where a harmless species mimics the warning signals of a harmful species, the adaptation being directed at one or more common predators

Müllerian mimicry
a situation where two or more harmful species (which may or may not be closely related and which share one or more common predators) mimic each other's warning signals

Fig. 4.10 Criocerine larva with faecal shield.

Fig. 4.11 Cassidine faecal shield.

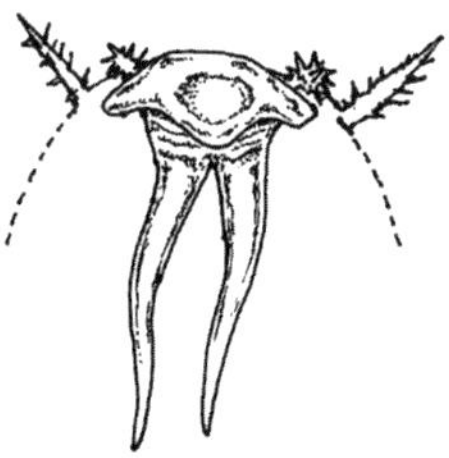

Fig. 4.12 Caudal fork of a cassidine larva.

been proposed, such as in the classic work by Wickler (1968). For example, *Clytra quadripunctata* and the scarce 7-spot ladybird *Coccinella magnifica* are both unpalatable Müllerian co-mimics (Donisthorpe, 1902) with black spots on an orange-red background. Colours may also be used defensively by groups as well as individuals. In the poplar-feeding species *Phratora laticollis*, the second and third larval instars occur in white and black forms (morphs). When feeding in groups, the mixture of colour morphs is less conspicuous than a uniformly coloured mass and may be seen less readily by predators or parasitoids (Grégoire, 1988). Adult colours may also provide camouflage against a background of host plant structures, such as the leaf-green dorsal surface of *Cassida viridis*.

Regurgitated or faecal matter may also form defensive structures. For example, *Crioceris* larvae coat their abdomens with regurgitated material, *Oulema* and *Lilioceris* carry a viscous faecal shield on their backs (Fig. 4.10) and *Cassida* carries a similar, but dry, shield attached to the caudal fork (Figs. 4.11, 4.12). Faecal shields hide or camouflage eggs and larvae (e.g. as droppings, small slugs, seeds or other plant structures) and create a physical barrier against predators and parasitoids which may include repellent secretions to deter ants and other enemies, as seen in *Altica* (Selman, 1988). Similarly, some larvae are protected dorsally by fragments of plant debris or the shed 'skins' (exuviae) from earlier instars, which in criocerine larvae bear a large droplet of plant and faecal fluids with embedded faecal pellets. Shields can protect eggs and larvae from poor conditions such as desiccation. The eggs of some species require maternal faecal coverings in order to end diapause and initiate embryonic development (Müller & Hilker, 2004). The African flea beetle *Weiseana berkeri* is one of the few confirmed examples of this (Nahrung & Marohasy, 1997). There are, however, costs associated with faecal shields, for example the ichneumon parasitoid wasp *Lemophagus pulcher* uses the shield of *L. lilii* to locate it as a host, probably via chemical and visual cues (Schaffner & Müller, 2001).

Chrysomelid eggs are often laid in exposed positions, but are glued firmly to prevent removal by predators, while some are laid in slits in plant stems, embedded in faeces or other materials (e.g. *Timarcha*) or laid in a hard case (ootheca) as in the cassidines. Some non-British genera such as *Chelymorpha* and *Ogdoecosta* attach their eggs to the ends of filaments, which may reduce the chance of encounters with predators and parasitoids (Olmstead, 1996). The larvae

diapause
a delay in development resulting in a period of dormancy

of many species that feed in exposed positions bear stout spines, which are longest in those that feed alone and have less well developed chemical defence glands. Much like *Clytra quadripunctata*, mentioned above, cryptocephaline larvae curl their soft abdomens under their bodies and enclose them in a felt-like case of plant material and faeces; if threatened, the larva can pull its whole body inside as well as producing deterrent secretions. As the larva grows, the case is continuously added to, and pupation takes place within it.

Most chrysomelids pupate in the soil, encased in a cocoon (puparium) of cemented soil and faeces. Some *Chrysomela* species pupate within the cuticle of the final larval instar (with defence glands intact), while others are uncased but have toothed 'traps' (Fig. 4.13) formed by the margins of abdominal segments which can be brought rapidly together by flexing the body and are effective against small predators and parasites. Cassidine pupae are also uncased, but bear large, and sometimes elaborate, spines on the expanded pronotum and sides of the abdomen. These spines may also have sensory functions, leading to defensive movement of the faecal shield (Olmstead, 1996).

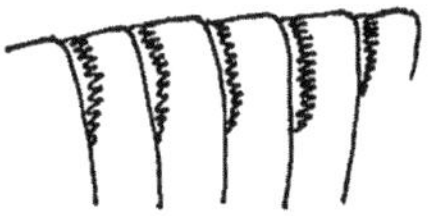

Fig. 4.13 Pupal trap.

Fig. 4.14 *Gastrophysa viridula* showing tarsal pads.

Fig. 4.15 *Timarcha* showing tarsal pads.

Adults of many species, especially 'typical' chrysomelines, are smooth and domed or rounded, making it difficult for small predators, such as ants, to grasp them. Similarly, cassidines are broadly flattened with splayed edges to the elytra and pronotum; as the head and appendages can be tucked into the underside, the beetle can adopt a limpet-like posture resistant to removal. The strength of grip is enhanced by the flat, felted tarsal pads (Figs. 4.14, 4.15) which attach to the substrate by surface tension when moistened by their secretions. With all stages from egg to adult being found on leaves, rather than hidden in the soil or plant tissues, cassidines display a wide range of well-developed defence mechanisms, and are a valuable group for study in this area.

The flea beetles (Alticinae) are named after their ability to jump in a flea-like manner and most species do so in order to avoid natural enemies as well as jumping as a mode of locomotion. They jump using greatly enlarged hind femora which contain an internal structure known as the 'metafemoral spring'. The spring is a modified tendon-like structure formed of proteins and fibres of chitin (Furth, 1988) and is curved and flanged to create a 'C' or 'S' shape, which allows the structure to be compressed and thus store energy. It does not contain the rubber-like protein resilin,

chitin
a polymer derived from glucose, forming much of the exoskeleton

which is used in the fleas' jumping apparatus, but it still allows flea beetles to jump at least as far as fleas (further in most cases) relative to body-length. Schmitt (2004) tested the jumping abilities of eight species and found that *Longitarsus anchusae* (body length 1.4–2.3 mm) could jump up to 289 times its own length. Relatively longer jumps were associated with shorter bodies and higher metafemoral muscle volume; in *Longitarsus*, jumping was also enhanced by elongated hind legs, and in *Psylliodes* by the highly compact form of the metafemoral spring. It is likely that during pre-jump crouching, flea beetles contract their tibia extensor muscles, storing elastic strain energy which is then released in a short burst using the triangular 'lever organ' as a trigger, although the exact mechanism of storage and release remains unknown.

5 Distribution and abundance

5.1 Biological aspects

In most species, even common and familiar ones, there is little information about aspects such as the number of eggs in a batch and lifetime female fecundity, and careful observations could yield valuable results. Where such information exists, there is considerable variation between and within species. For example, in *Donacia clavipes* eggs are laid in rows of about 25 along the edges of leaves. Similar numbers are seen in several other species of *Donacia*, while gravid *Plateumaris affinis* and *Plateumaris discolor* have been found containing around 50 eggs. In *Zeugophora subspinosa*, eggs are laid singly or in pairs with up to six per leaf. It is uncertain how many a female may lay in total. Where eggs are laid in pairs and both hatch, only one larva appears to survive, and the cause of this is also uncertain. Observations could help clarify female fecundity and whether there is direct competition between pairs of larvae. In *Crioceris asparagi*, eggs are often laid in lines of three to eight along asparagus stems (Fig. 5.1) and *Lilioceris lilii* lay a similar number in irregular groups. Some species do not lay in any discernible pattern, for example *Cryptocephalus coryli* lays up to 10 eggs per day singly and apparently at random for a period of 10 days to a month. One of the most familiar British species, the green dock beetle *Gastrophysa viridula* known for the swollen black abdomens of gravid females and its abundance on dock plants, lays larger batches of about 45 on the underside of the leaves. Unlike some species such as the relatively large *Chrysomela* and *Timarcha*, which may survive for two years, *G. viridula* adults live for about three weeks but lay eggs more-or-less continuously for the last two weeks. However, the total fecundity of *G. viridula* is not known, nor how it compares with that of longer-lived, but possibly less intensively egg-laying species. In *Oomorphus concolor*, with up to 13 eggs laid over a week, total fecundity may be considerably lower than in *G. viridula*. There is no single, readily accessible reference work giving details of fecundity for British species. This would be a valuable project and would also highlight species where research could fill gaps in knowledge about the fecundity of individual species.

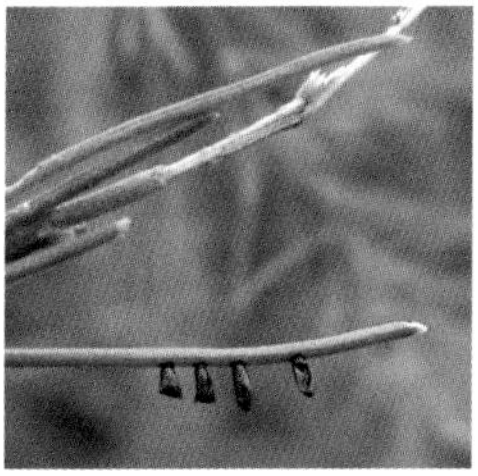

Fig. 5.1 Asparagus beetle eggs.

Egg-laying is also affected by environmental conditions. In *Galerucella pusilla*, eggs are laid on plants in small batches, and the total number laid per day falls from 10 or more with summer temperatures of 25–30°C to only a few eggs per day when cooled below 20°C. In a few cases, total fecundity is known. For example *Cryptocephalus biguttatus* can produce approximately 170 eggs in one month, *Cryptocephalus bipunctatus* can lay over 300 in two weeks and *Cryptocephalus fulvus* lays around 35 eggs in a month. Similar levels of variation can also be seen between individuals of a species, such as *Cryptocephalus nitidulus* which Owen (2003) showed could lay 103–312 eggs (average 180) over 19–45 days (average 30).

diapause
a delay in development resulting in a period of dormancy

Timarcha goettingensis lays eggs singly or in groups of up to four, with a total of 50–100 being produced by each female. In this species, the oviposition period is split in two by hibernation, and the autumn eggs undergo diapause early in their development due to low winter temperature, while those laid in spring do not. *Timarcha tenebricosa* eggs occur in slightly larger groups of up to eight and also undergo diapause, though at a later stage as a fully developed embryo. In both species, eggs are coated in regurgitated food material on plant stems or just beneath the surface of the soil.

Palaearctic
an ecozone (a large area where species have a long shared evolutionary history) covering Europe, Asia north of the Himalayas, north Africa and the northern and central parts of the Arabian Peninsula

Some species of *Chrysolina*, such as *Chrysolina staphylaea* and *Chrysolina varians*, may either lay eggs or show viviparity and give birth to live young. The difference in strategy may be related to latitude (i.e. temperature and the period of favourable conditions during the year) and it is possible that *C. staphylaea* only reproduces viviparously in the northern part of its Palaearctic range. Total fecundity is likely to be lower in viviparous species such as *Gonioctena* as more resources are put into producing fully developed larvae, but again there is little information on the numbers of larvae deposited, and there can be geographical differences as there are for the reproductive strategy itself. For example, *Gonioctena decemnotata* produces around 40 larvae in southern Europe, but the figure in Britain may be different.

As well as fecundity, population distribution is affected by dispersal ability and therefore whether or not a species has functional wings and the ability to fly. This affects not only a species' ability to disperse within Britain but also to colonise, or recolonise, across stretches of sea. Observations have been made of very local dispersal by flight, such as crossing a few metres between food plants, and occasionally longer flights of up to approximately 1.5 km (in *Lochmaea caprea*). Few measurements of longer-distance flying ability have been made, although Palmén (1944) reported mass dispersal of *Chrysomela aenea* in northern Europe. One of the few species where this has been researched more thoroughly is the Colorado potato beetle *Leptinotarsa decemlineata* due to its economic importance as a pest. This was found to make short flights (to find foodplants, mates and egg-laying sites), medium-length flights to hibernation sites, and long-distance migration (Cox, 2004). In most cases in Britain, mark-and-recapture experiments, or an equivalent approach, would be needed to, for example, investigate movements within the country or to discover how readily leaf beetles can cross the English Channel. Note that variation in wing development is covered in Section 3.3.

5.2 Distribution

In most cases, population sizes are unknown, even as rough estimates. There are a very few exceptions such as the Lundy cabbage flea beetle *Psylliodes luridipennis* which, as an endemic species with a highly restricted distribution, is subject to census-type surveys which aim to derive population estimates.

Others such as the Pashford pot beetle *Cryptocephalus exiguus* illustrate the need for landscape-scale conservation efforts. Its common name comes from the only site with recent records, Pashford Poors Fen in Suffolk, although it was previously known from a small number of scattered sites in eastern England and Somerset. Following a decline due to drainage, wider water abstraction, meadow improvement and conversion to arable use, it became restricted to this single site. However, although listed as a UK Biodiversity Action Plan (BAP) species, with Pashford Poors Fen designated a Site of Special Scientific Interest (SSSI) and Suffolk Wildlife Trust reserve, habitat degradation was caused by borehole abstraction from adjacent intensively managed land. It requires damp conditions with a high water table, and suitable habitat no longer exists at Pashford Poors Fen. As targeted surveys have not found the beetle there since 2000, it is likely to be extinct in Britain unless found to be present at undiscovered sites.

Some species may colonise naturally from outside Britain, while others expand or decline in range along with their foodplants, either wild or cultivated, or in response to climate change. Some may be accidentally introduced with imported foods (as is the case with a number of bruchids) or garden plants, and others may, if necessary, be the subject of reintroduction projects. Whatever the causes of change, species records are needed to follow shifts in distribution, and recording may target a particular species or be more general in scope.

It is beyond the scope of this book to give details of the distribution of all British species, however, the vast majority are mapped in Cox (2007) with few significant changes since then. Here, a selection is included to illustrate different aspects of distribution, such as those that have declined or spread in recent years, or that have a restricted distribution such as montane species. The most familiar species are also included. Others are of economic importance or show the effects of changing land use such as the increase or decline in particular crops. This also provides an opportunity to include up-to-date data collected since the publication of Cox's Atlas. Some are included because their distribution, biology or ecology are poorly understood and targeted observations could provide notable improvements to knowledge of the species. For listings of scarcity and rarity categories of British species, see the status review by Hubble (2014), which is available as a free download.

5.3 Development of the British fauna

Several fossil or subfossil chrysomelids are known from Britain, with the following recovered from Quaternary sediments between 10,000 and 130,000+ years old (Duff, 2014):

Chaetocnema obesa (Boieldieu)
Chrysolina limbata (Fabricius)
Chrysomela collaris Linnaeus
Chrysomela septentrionalis (Ménétries)
Donacia polita Kunze
Entomoscelis adonidis (Pallas)
Neocrepidodera interpunctata (Motschulsky)
Pachnephorus tesselatus (Duftschmid)

The current British fauna consists of post-glacial survivors plus those species that colonised after ice retreat. This process continues. It is augmented by accidental introductions due to human activity and also by colonisation aided by the increasing temperatures associated with climate change. Like other species groups, the British fauna remains impoverished to some extent by glaciation with 128 species of Alticini out of around 280 species of chrysomelids, while for example the Czech Republic and Slovakia combined have 235 species of Alticini alone (Čižek & Doguet, 2008) and France has around 300 (Doguet, 1994).

A small number of species may be relicts of a colder Ice Age climate, such as *Phratora polaris* which is restricted to a few areas of limestone grassland between 700 m and 1,100 m in altitude in the mountains of north-west Scotland. Outside Britain it is found across the northern Palaearctic from Icelandic heaths to Kamchatka, feeding on alpine willows. Warming is therefore likely to impact this species by forcing it to higher altitudes and thus smaller areas, although it is probably under-recorded given the inaccessibility of its habitat.

A greater number of species are accidental introductions. Some are well known (if unpopular) such as the lily beetle *Lilioceris lilii*, which arrived with imported ornamental plants. Other less familiar arrivals include several bruchines associated with dried foods such as legumes, and the newly described Tasmanian species *Paropsisterna selmani* which feeds on eucalypts and has also been found in Ireland, again travelling with imported ornamentals (Reid & de Little, 2013). With some new discoveries such as *Longitarsus obliteratoides* (see below), it can be less clear whether the species is a recent colonist or has simply been overlooked. Many accidental imports are recorded singly or sporadically but do not colonise outside of heated buildings or storage facilities, often because they are unable to complete their life cycle due to cold winter temperatures or because their foodplant does not grow in Britain. Examples include several of the bruchines mentioned above such as *Bruchus lentis* and *Callosobruchus rhodesianus*. In some cases where pest status is likely, eradication measures are in place, as for the infamous Colorado potato beetle.

5.4 Selected distribution maps

These are given in taxonomic order to match the species list given in Section 6.3. Body lengths are given as ranges excluding appendages (legs, antennae and mouthparts).

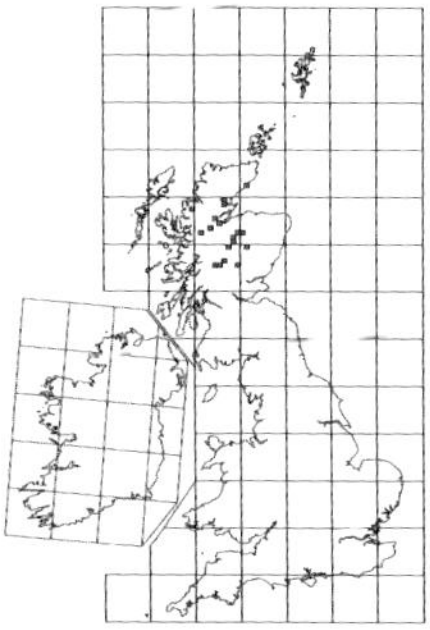

Map 1 *Zeugophora turneri.*

Zeugophora turneri

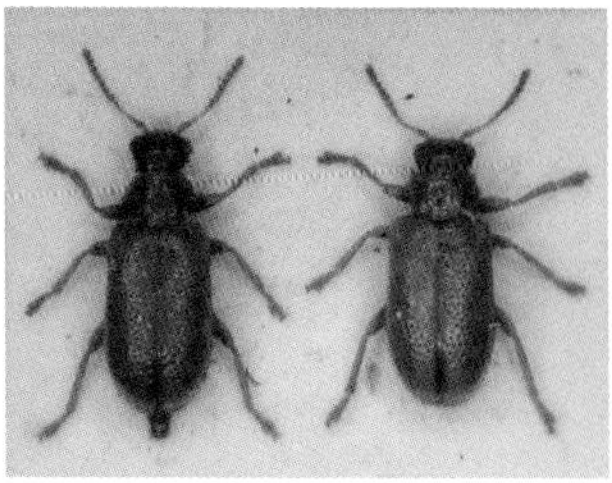

A Palaearctic species. In Britain, found only in central and northern Scotland where it is restricted to a small number of sites having declined and been lost from some locations. Adults feed on young birch and aspen, with larvae as leaf miners on aspen. 3.2–3.6 mm.

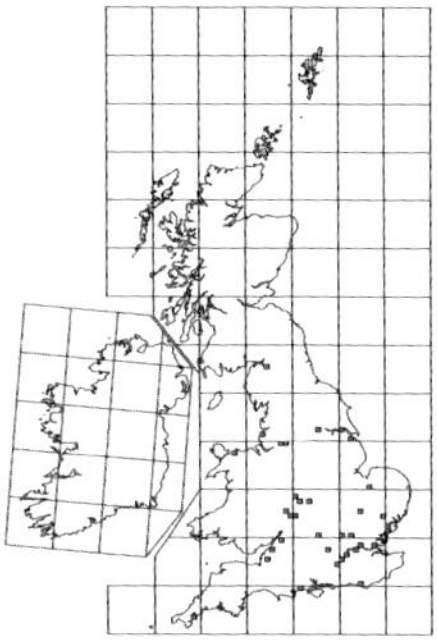

Map 2 Pea beetle *Bruchus pisorum.*

Bruchus pisorum

Pea beetle. Widespread but scattered as most records are from facilities where peas are stored, or are escapes from such locations. Though likely to have originated in the eastern Mediterranean region it is now found worldwide due to being transported with peas. 3.4–4.5 mm.

Map 3 *Macroplea appendiculata.*

Macroplea appendiculata

A Palaearctic species. Widespread but sparsely scattered throughout Britain, having declined with populations lost from some sites such as Loch Leven. Found on a variety of aquatic and marginal plants in rivers, lakes, canals and drainage ditches. 6.0–7.5 mm.

Map 4 *Macroplea mutica.*

Macroplea mutica

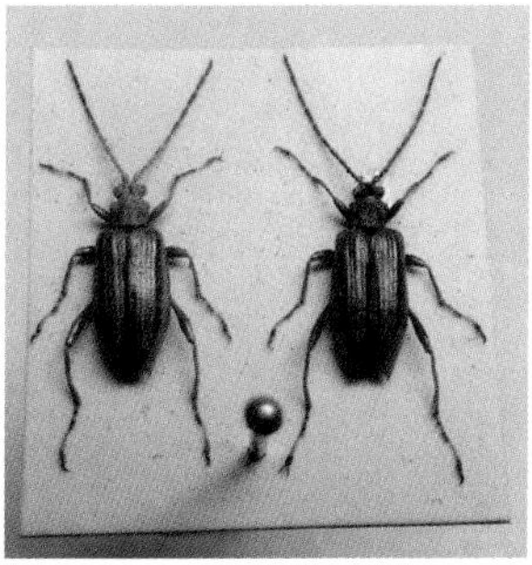

A Palaearctic species. Rare in Britain, previously sparsely scattered from south-east to north-west England but following a marked decline, now mainly in the east and south-east. Found on various plants (especially fennel pondweed *Potamogeton pectinatus*) in brackish water, usually coastal clay pits and dykes, though sometimes associated with inland saline lagoons. Adults feed on submerged leaves, with larvae on the roots. 5.0–7.0 mm.

Map 5 Two-tone reed beetle *Donacia bicolora.*

Donacia bicolora

Two-tone reed beetle. A Palaearctic species. Widely scattered in southern Britain with clusters in the heathlands of Surrey and Berkshire, also around the lakes of Fermanagh, Northern Ireland. Adults found by flowing water, sometimes standing water, mainly on the leaves of branched bur-reed *Sparganium erectum*. Larvae are found on the rhizomes of this plant with cocoons also known from the roots of common club-rush *Schoenoplectus lacustris*. 8.0–10.2 mm.

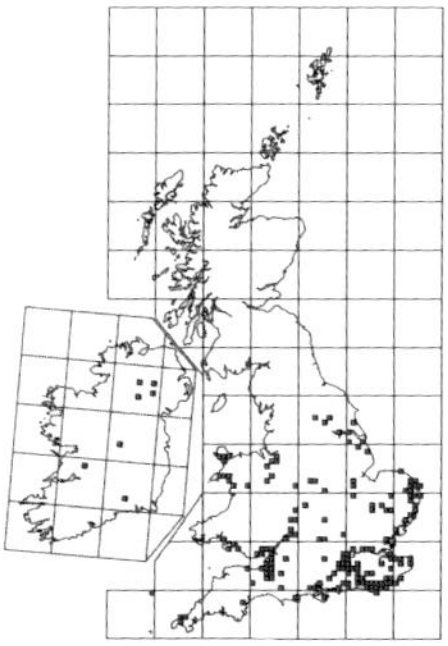

Map 6 *Donacia marginata.*

Donacia marginata

A Palaearctic species. Widespread and locally common, especially in southern England. Found in wetlands and around standing water on emergent vegetation such as *Phragmites* reed-beds. 8.0–10.0 mm.

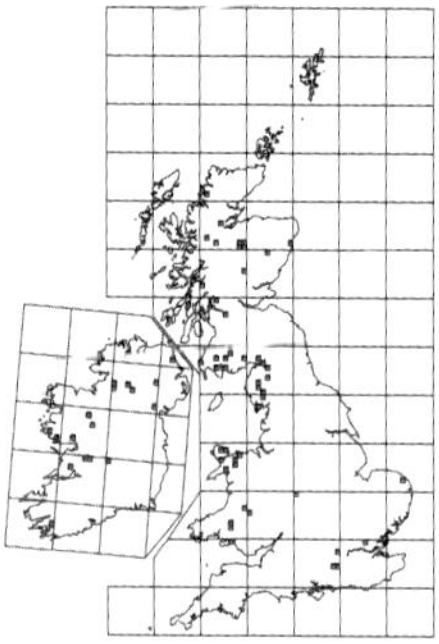

Map 7 *Donacia obscura*.

Donacia obscura

A Palaearctic species. Widely scattered, mostly in the north and west of Britain and Ireland, in boggy habitats by standing and running water, and in acid pools near the coast. The adults are found on a range of Cyperaceae such as club-rushes and sedges in locations where this family dominates the vegetation. Larval habits are unknown. 8.1–10.7 mm.

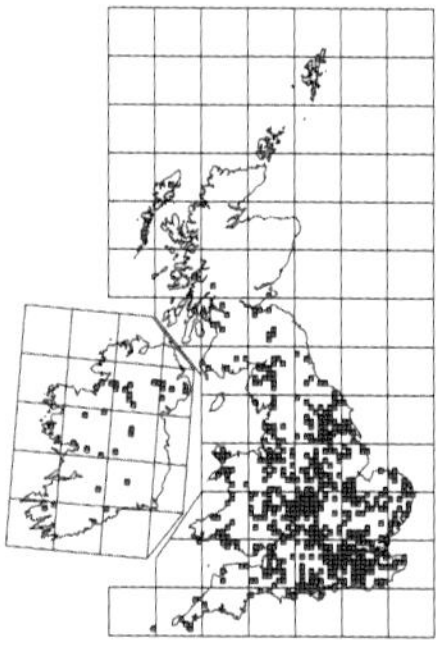

Map 8 *Donacia simplex*.

Donacia simplex

A Palaearctic species. Common and widespread in Britain, especially in England and Wales. Adults are found mainly on *Sparganium* bur-reeds in a variety of aquatic habitats, with larvae at the roots. 5.4–9.4 mm.

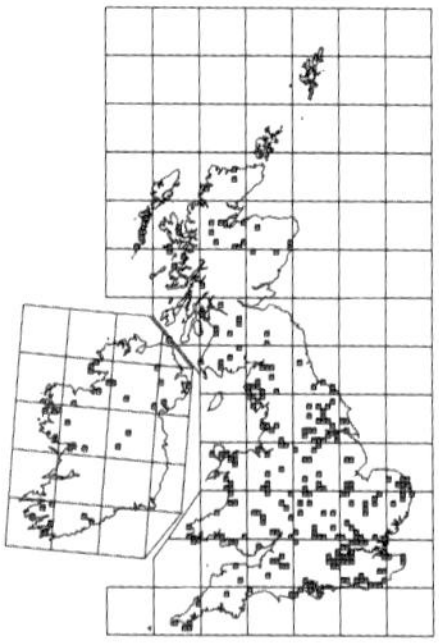

Map 9 *Donacia versicolorea*.

Donacia versicolorea

A western Palaearctic species. Widespread but not common throughout Britain. Adults and larvae are found mainly on *Potamogeton* pondweeds in a variety of aquatic and wetland habitats. 6.2–8.9 mm.

Map 10 *Oulema erichsoni.*

Oulema erichsoni

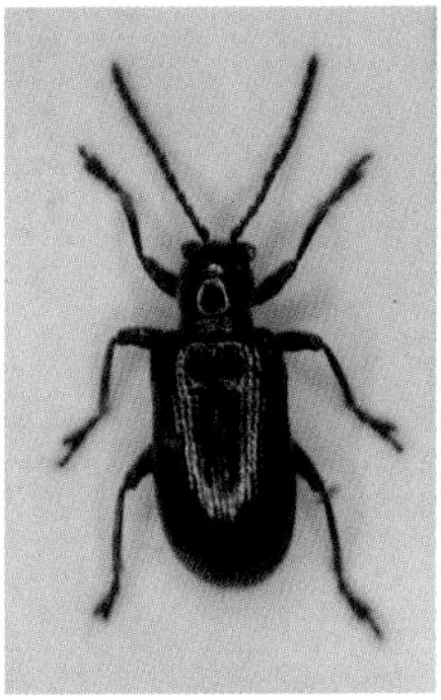

A Palaearctic species. In Britain, there are recent records only from Somerset. Adults and larvae feed on the leaves of floating sweet-grass *Glyceria fluitans*, mainly in wet peat cuttings or trenches with little other vegetation, or on heaths. There is little suitable habitat due to losses through drainage and drying of the cut peat surface. Severe floods during the winter of 2013/14 may have impacted the entire extended population within the Somerset Levels, but this is as yet uncertain. 4.0–4.5 mm.

Map 11 *Oulema septentrionis.*

Oulema septentrionis

A western Palaearctic species. In the British Isles, restricted to Ireland where it is found in a range of habitats, mainly near water, including coastal areas. Adults and larvae are mainly on oats, but may also be found on bulrush fronds. 4.0–4.5 mm.

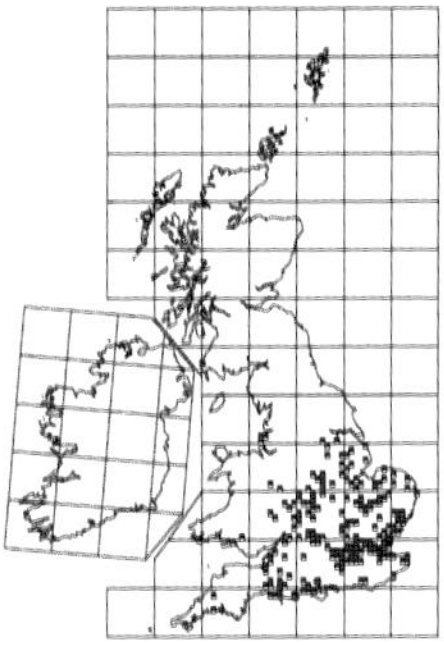

Map 12 Asparagus beetle *Crioceris asparagi.*

Crioceris asparagi

Asparagus beetle. A western Palaearctic species. In Britain, widespread and sometimes common in England, especially in the south and east. Adults and larvae feed on asparagus leaves and may be considered a pest of commercial and garden plants. 5.0–6.5 mm.

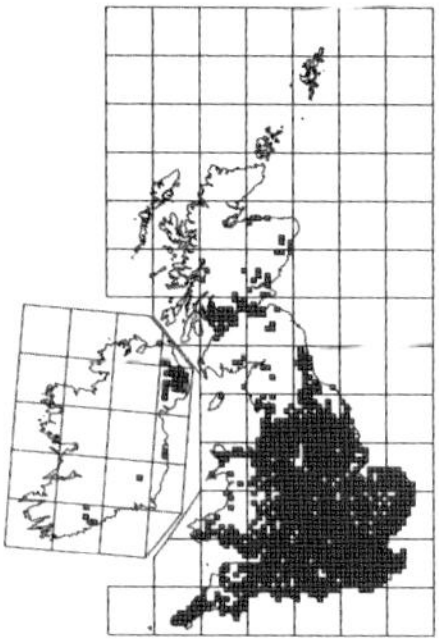

Map 13 Lily beetle *Lilioceris lilii.*

Lilioceris lilii

Lily beetle. An Asian species first recorded in Britain in the 19th century. It began a rapid and ongoing range expansion in the 1980s and is now found throughout much of the country, especially southern and eastern England. Adults and larvae feed on various members of the lily family, mainly in gardens and nurseries where they may be considered a pest, occasionally in natural habitats. 6.0–8.0 mm.

Map 14 Hazel pot beetle *Cryptocephalus coryli.*

Cryptocephalus coryli

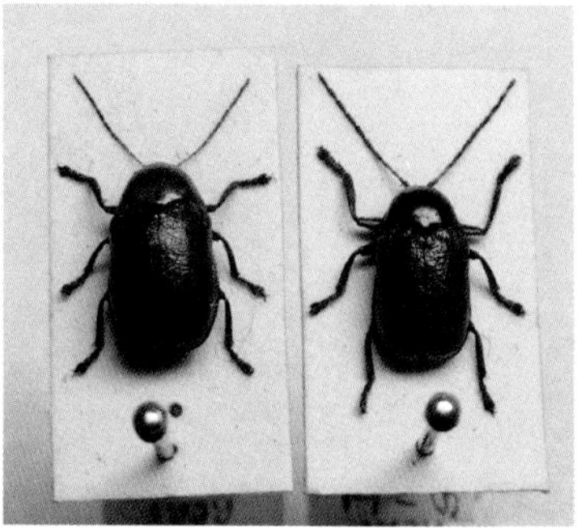

Hazel pot beetle. A Palaearctic species. In Britain, once widely scattered from southern England to Scotland but now restricted to a few sites in southern and central-eastern England following a serious decline since the 1950s. This was due to habitat loss through clear-felling and conversion to conifer forestry, as well as habitat neglect leading to development into high forest. There is ongoing survey work in Sherwood Forest, and recent records from a release site in Lincolnshire. Tree-top surveys have provided useful records, so it is possible that it exists at other locations but has been overlooked when searching at ground-level. It is found in clearing and ride margins in broadleaved woodland on south-facing slopes, chalk downland, and on moors and heathland. Adults feed on leaves of birches and other trees. Larvae feed on fallen leaves and possibly fallen catkins. 5.8–7.5 mm.

epicormic
developing from previously dormant buds hidden beneath the bark of a trunk or branch in woody plants

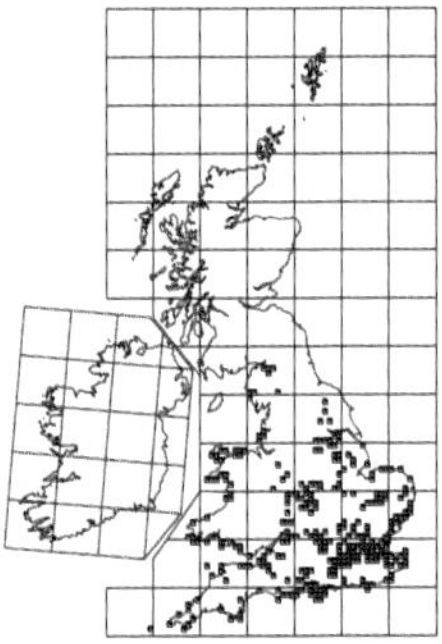

Map 15 *Cryptocephalus fulvus.*

Cryptocephalus fulvus

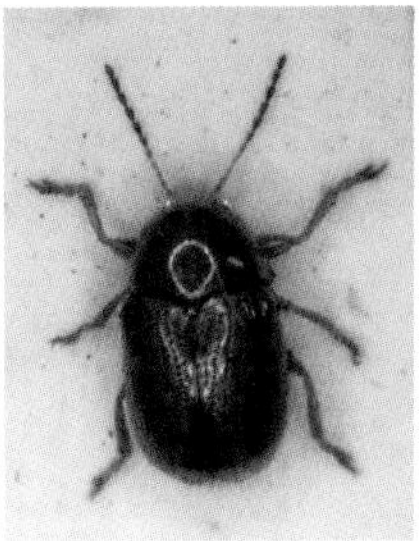

A Palaearctic species. In Britain, widespread in England and Wales and mainly coastal in the west. Found on several plant species in a range of mostly open habitats. 2.0–3.0 mm.

Map 16 *Cryptocephalus hypochaeridis.*

Cryptocephalus hypochaeridis

A Palaearctic species. In Britain, scattered in open habitats in England and Wales with clusters on the North Downs and some other areas of limestone and western dune systems. The life cycle has not been described but adults are associated with yellow flowers such as hawkweeds and hawkbits. 4.6–5.7 mm.

Map 17 Oak pot beetle *Cryptocephalus querceti.*

Cryptocephalus querceti

Oak pot beetle. A Palaearctic species. In Britain, previously scattered in England with records as far north as Lancashire. Now declined to three sites; Windsor Great Park (Berkshire), Donington Park (Leicestershire) and Sherwood Forest (Nottinghamshire). It is associated mainly with mature oaks, but occasionally also hawthorn and possibly birches. Its habitats are ancient broadleaved pasture-woodland, parkland and forests, favouring open parkland over woodland with a closed canopy. Adults feed on oak leaves, preferably fresh and tender ones, and have been found on epicormic growth, while larvae feed on debris such as oak litter in holes within the trunk. 2.5–3.4 mm.

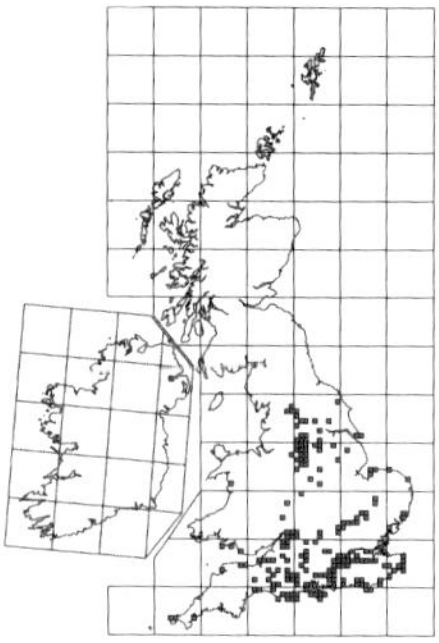

Map 18 Lesser bloody-nosed beetle *Timarcha goettingensis*.

Timarcha goettingensis

Lesser bloody-nosed beetle. A western Palaearctic species. In Britain, widespread but patchy as far north as Yorkshire, with a small number of records from Wales and Ireland. Adults and larvae are found in various open habitats where they feed on bedstraws, and are also known from crosswort and woodruffs. 8.0–13.0 mm.

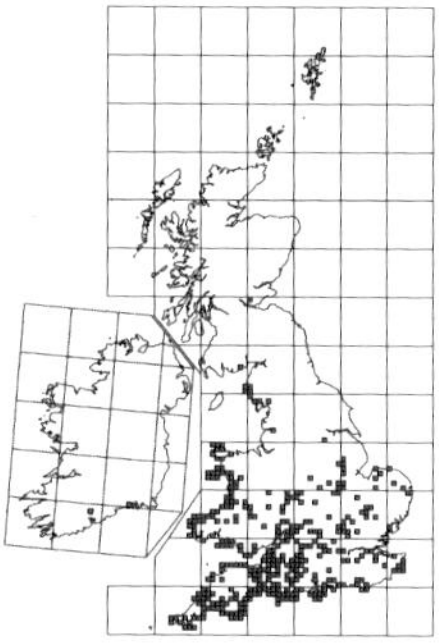

Map 19 Bloody-nosed beetle *Timarcha tenebricosa*.

Timarcha tenebricosa

Bloody-nosed beetle. A Palaearctic species. Widespread in the southern half of Britain with scattered, mainly coastal, records further north. Records also tend to be near the coast in west Wales and south-west England. Adults and larvae are found in various open habitats on well-drained soil where they feed on bedstraws, and are also known from crosswort, madders and woodruffs. 11.0–18.0 mm.

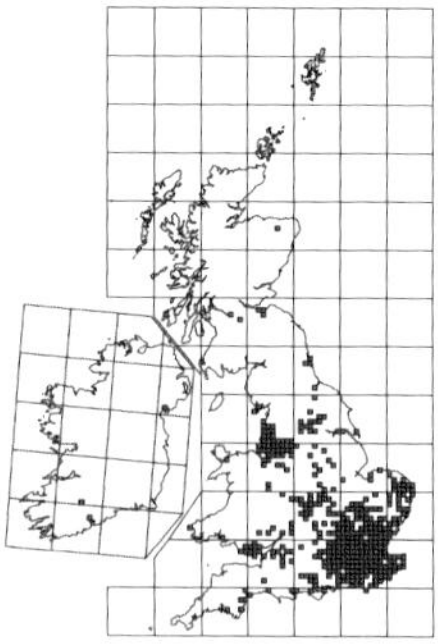

Map 20 Rosemary beetle *Chrysolina americana*.

Chrysolina americana

Rosemary beetle. A Mediterranean species. First recorded in Britain in 1963, now widespread after a large range expansion, though only scattered records from Scotland and Ireland. Adults and larvae feed mainly on rosemary and lavender, though they are occasionally found on other plants such as sage and thyme. 6.7–8.1 mm.

Map 21 *Chrysolina banksi*.

Chrysolina banksi

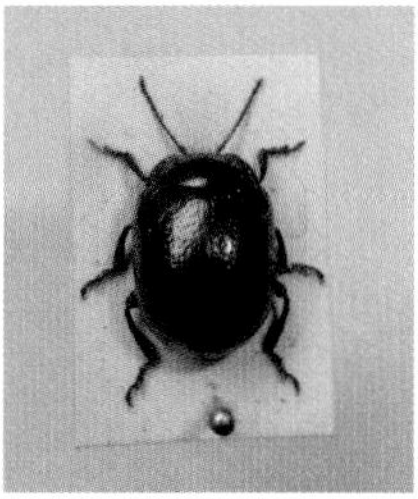

A mainly Mediterranean species also known from the Atlantic coast of various countries, Atlantic islands and parts of Asia. In Britain, widespread and usually coastal, particularly in south and south-west England, west Wales and the Isle of Man. Found on a wide range of plants in open habitats, and can be locally common. 8.0–10.7 mm.

Map 22 Rainbow leaf beetle *Chrysolina cerealis*.

Chrysolina cerealis

Rainbow leaf beetle. A Palaearctic species. In Britain, known only from a small number of sites in Snowdonia where it feeds on thyme in mountain grasslands. 7.0–8.0 mm.

Map 23 Tansy beetle *Chrysolina graminis*.

Chrysolina graminis

Tansy beetle. A Palaearctic species. In Britain, following a marked decline and reduction in range, it is currently confirmed in any numbers only from the area around York where there is a series of sub-populations along approximately 45 km of the River Ouse. Records in some other locations are dubious due to confusion with *Chrysolina herbacea* e.g. a large population supposedly on the River Trent was reported in error. Severe floods during, and since, the winter of 2013/14 may have impacted the entire extended population along the Ouse.

Although hibernating adults can survive winter inundation, the extent and duration of flooding was greater than usual and it is possible the whole national population was impacted to some extent. However, survey work is required to confirm this. 7.7–10.5 mm.

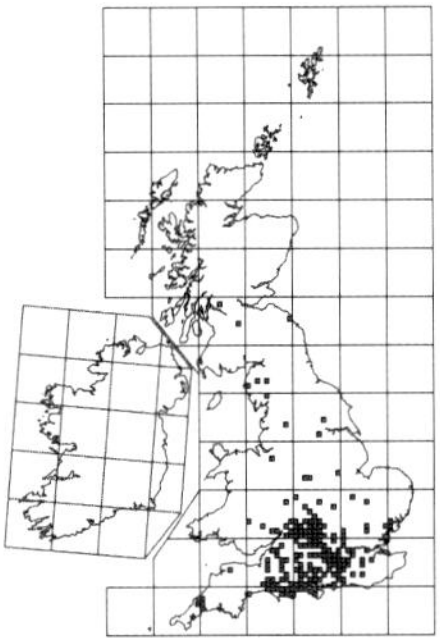

Map 24 Mint leaf beetle *Chrysolina herbacea.*

Chrysolina herbacea

Mint leaf beetle. A Palaearctic species, including some Mediterranean islands, Iraq and also the Indian subcontinent. In Britain, mainly in central southern England, but with scattered records north into Scotland. It is found mainly in wet habitats, as well as grasslands and gardens, and feeds on mints, especially water mint. 7.0–11.0 mm.

Map 25 *Chrysolina latecincta.*

Chrysolina latecincta

Known only from Scotland and Norway. In Scotland, the only known sites are Orkney cliff-tops and a saltmarsh at Loch Etive, Argyll. It is found in grassy, salty cliff-top vegetation, on cliff edges with small patches of vegetation among bare earth and rocks, an old cliff-edge sandstone quarry and a saltmarsh. Adults and larvae feed on the leaves of several plant species including plantains, toadflaxes and snapdragon *Antirrhinum majus. Chrysolina intermedia* is now considered a subspecies of *C. latecincta*. 7.0–11.0 mm.

Holarctic
a large area consisting of the Nearctic (North America) and Palaearctic ecozones (see above).

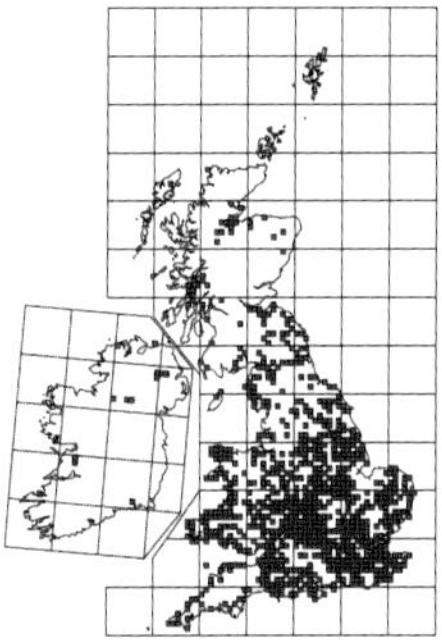

Map 26 Knotgrass leaf beetle *Chrysolina polita*.

Chrysolina polita

Knotgrass leaf beetle. A Palaearctic species, including parts of the Middle East. In Britain, widespread and common on various plants in a wide range of habitats. 5.9–8.6 mm.

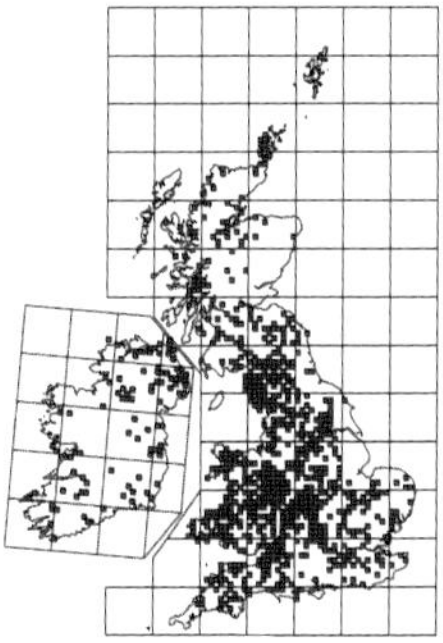

Map 27 Green dock beetle *Gastrophysa viridula*.

Gastrophysa viridula

Green dock beetle. A Holarctic species. In Britain, widespread and common on docks in many habitats where it is a familiar sight feeding on leaves, especially in the west of the country. It is occasionally found on various other plants, although it is uncertain whether the life cycles can be completed without docks. Feeding trials could be undertaken to clarify this on other members of the dock family (Polygonacaeae) as well as the wider range of plant families which may be eaten by this beetle. 4.0–6.0 mm.

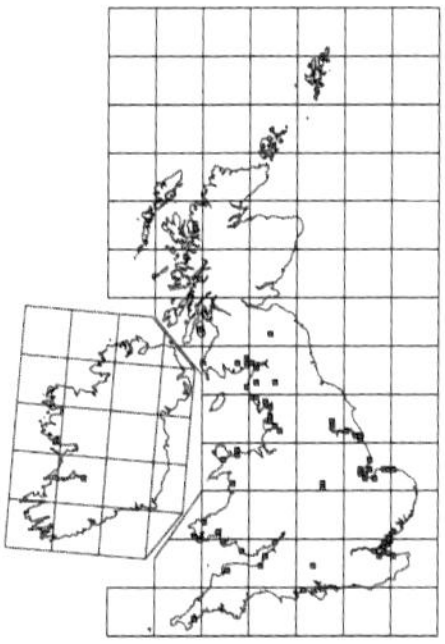

Map 28 *Phaedon concinnus*.

Phaedon concinnus

A Palaearctic species. In Britain, scattered around the coast in a variety of habitats where it feeds on sea plantain *Plantago maritima* and sea arrowgrass *Triglochin maritimum*, although it is possible only the adults eat the plantain. 3.2–4.1 mm.

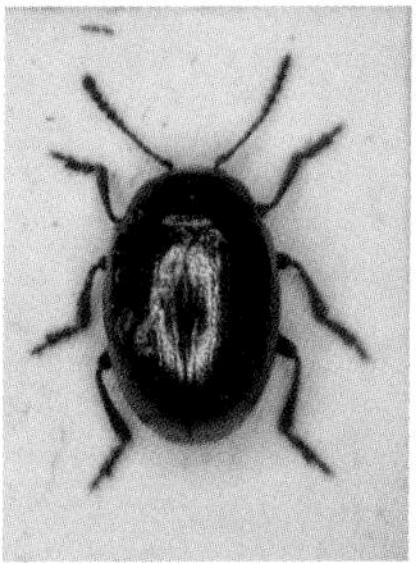

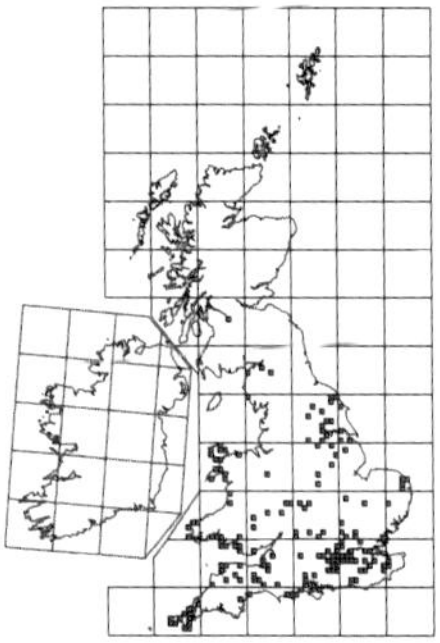

Map 29 Red poplar leaf beetle *Chrysomela populi.*

Chrysomela populi

Red poplar leaf beetle. A Palaearctic species. In Britain, widespread but patchily scattered in England and Wales. Found in various habitats where adults and larvae feed on willows and poplar saplings. 10.0–12.0 mm.

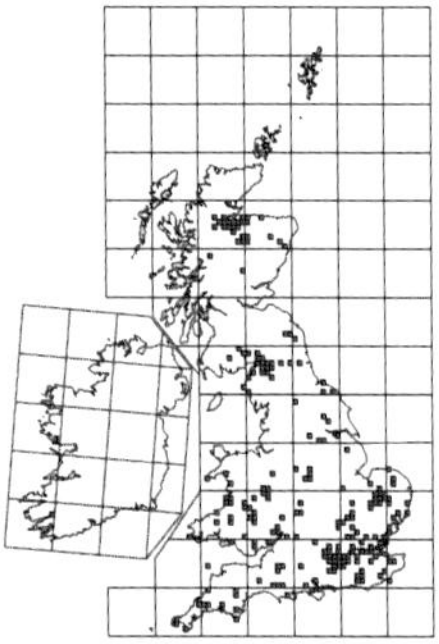

Map 30 Broom leaf beetle *Gonioctena olivacea.*

Gonioctena olivacea

Broom leaf beetle. A western Palaearctic species, including parts of North Africa. In Britain, widespread but patchy in a range of habitats. They feed mainly on brooms, but adults also feed on lupins and may be found on gorse and related plants, while larvae also feed on Lundy cabbage *Coincya wrightii*. 3.7–5.2 mm.

Map 31 *Phratora polaris.*

Phratora polaris

A north Eurasian species. In Britain, it occurs only at altitudes between 700 and 1,100 m on mountains in north and west Scotland, in a few areas of grassland on dolomitic limestone outcrops where shoots of dwarf willow *Salix herbacea* wind through *Racomitrium lanuginosum* moss. It is found under stones among either of these plants, and both adults and larvae feed on dwarf willow leaves. 3.8–4.6 mm.

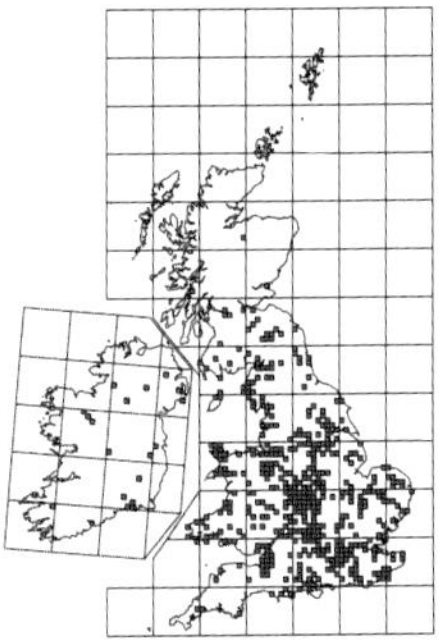

Map 32 *Galerucella nymphaeae/sagittariae* complex.

Galerucella nymphaeae/sagittariae complex

A Holarctic species complex. See Hubble (2012) for more about separating these very similar species which have few, if any, reliable external features to distinguish them. In Britain, widespread and common in various habitats, both wetland and drier areas. In aquatic habitats, they are found on water-lilies. In semi-aquatic and drier habitats, they are found on a range of plants, especially docks and Rosaceae. 4.0–8.0 mm.

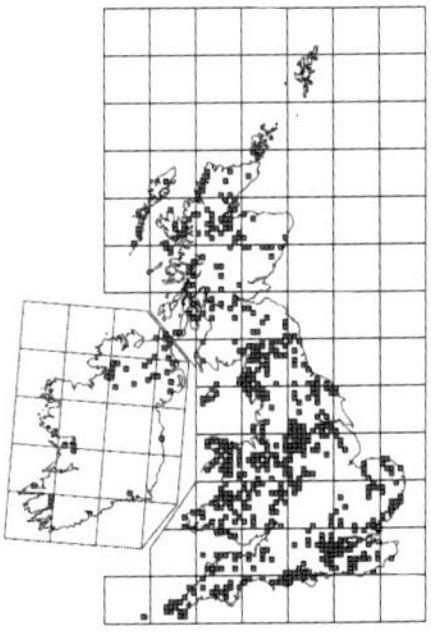

Map 33 Heather beetle *Lochmaea suturalis.*

Lochmaea suturalis

Heather beetle. A Palaearctic species. In Britain, widespread and found wherever there is heather *Calluna vulgaris*, or occasionally heaths (*Erica* species), usually with a damp layer of moss or plant litter. 4.3–6.0 mm.

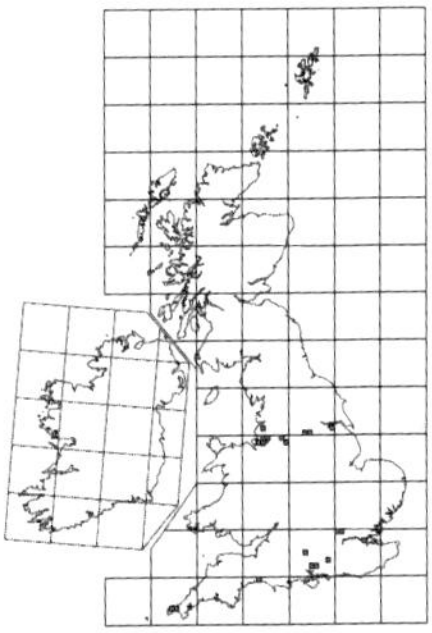

Map 34 Alder leaf beetle *Agelastica alni.*

Agelastica alni

Alder leaf beetle. A Palaearctic species also introduced to North America. In Britain, its distribution is not well understood. It was previously thought to be extinct but since 2004, populations have been found in the Manchester, Sheffield and Southampton areas among others. It is found mainly on alders, though sometimes on other trees, in a range of habitats. It has generally been associated with wetlands, especially alder carr, but many recent records are from urban locations. 6.0–7.0 mm.

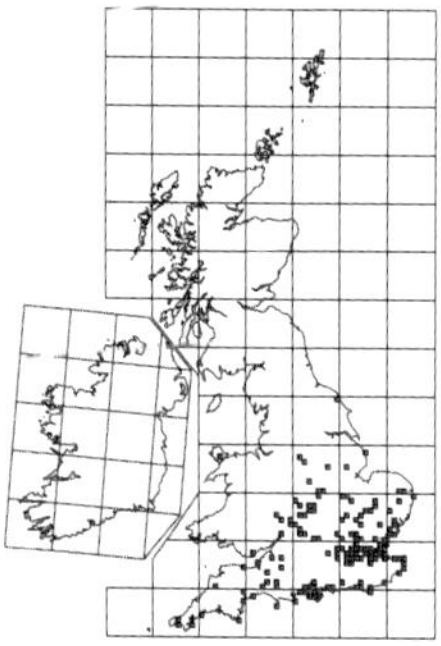

Map 35 Cabbage flea beetle *Phyllotreta cruciferae.*

Phyllotreta cruciferae

Cabbage flea beetle. A Palaearctic species also found in northern and eastern Africa, and introduced to North America. In Britain, widespread but patchy in England, mainly in the south and midlands. Found on a wide range of plants, mostly Brassicaceae including crops. 1.8–2.5 mm.

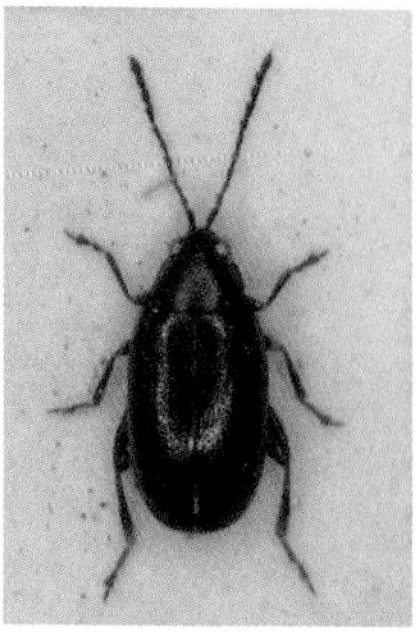

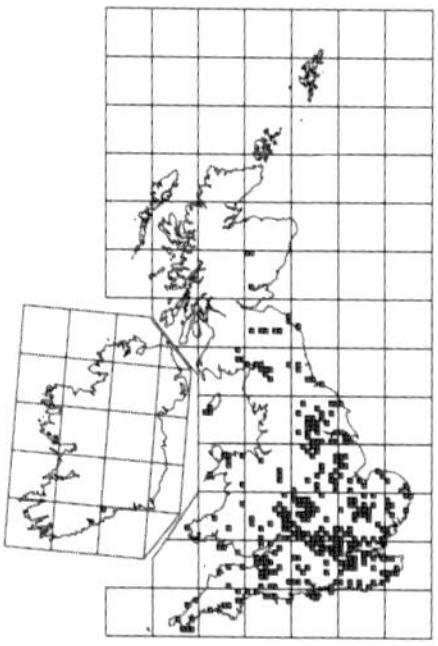

Map 36 Large striped flea beetle or turnip flea beetle *Phyllotreta nemorum.*

Phyllotreta nemorum

Large striped flea beetle or turnip flea beetle. A Palaearctic species, also introduced to Australia. In Britain, widespread in many habitats on a wide range of plants, mostly Brassicaceae including crops. 2.4–3.5 mm.

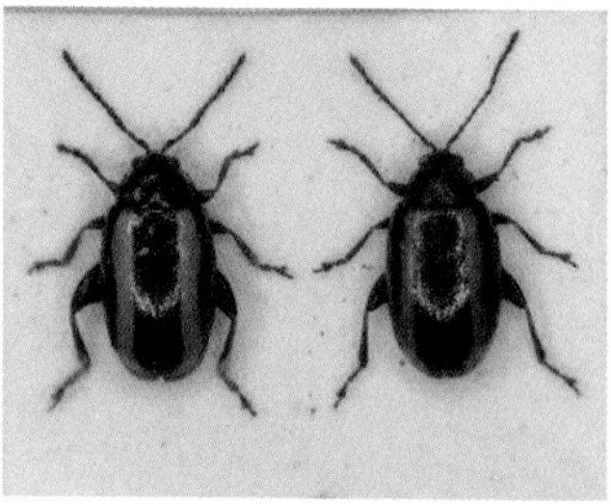

Map 37 Striped flea beetle *Phyllotreta striolata.*

Phyllotreta striolata

Striped flea beetle. A Palaearctic species, now found throughout Europe and Asia, and introduced to North America and South Africa. In Britain, sparsely scattered in England and Wales. Found in a range of habitats on various Brassicaceae including crops. 1.9–3.5 mm.

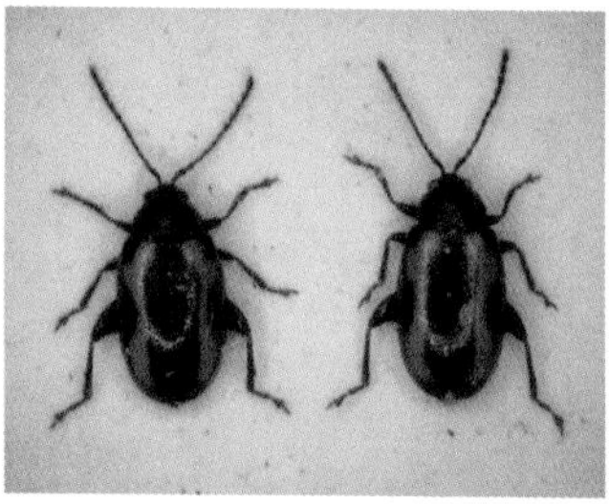

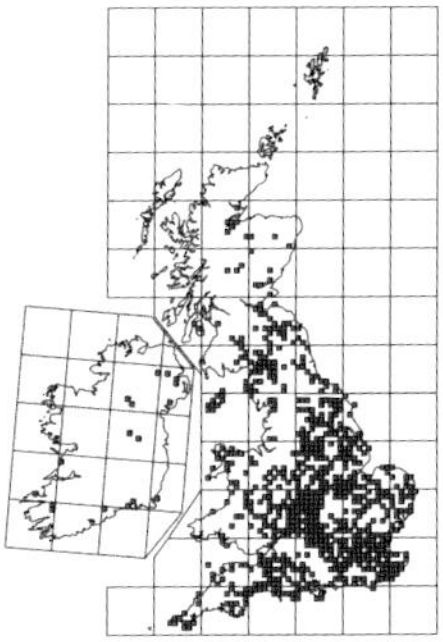

Map 38 Small striped flea beetle *Phyllotreta undulata*.

Phyllotreta undulata

Small striped flea beetle. A Palaearctic species, now found throughout Europe and central Asia, and introduced to North America, Oceania and Australasia. In Britain, widespread and common in many habitats on a wide range of plants, mostly Brassicaceae including crops. 2.0–2.8 mm.

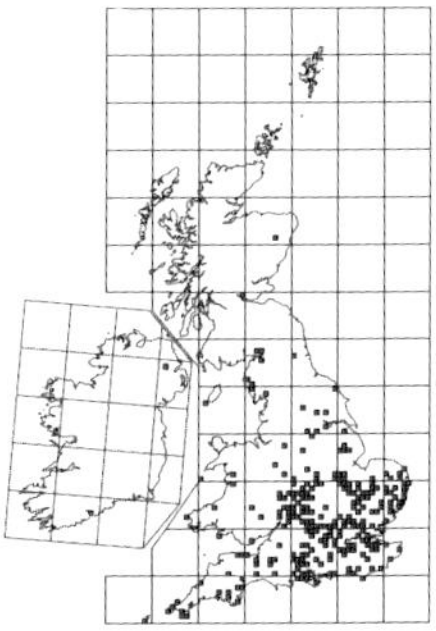

Map 39 Barley flea beetle *Phyllotreta vittula*.

Phyllotreta vittula

Barley flea beetle. A Palaearctic species, also in North Africa and introduced to North America. In Britain, widely scattered in southern England, with sparse records further north and in Wales. It is found in various habitats on a range of Brassicaceae and grasses, including cereals. 1.8–2.3 mm.

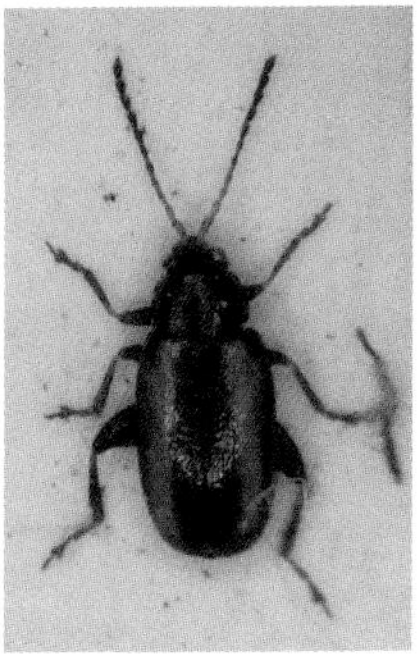

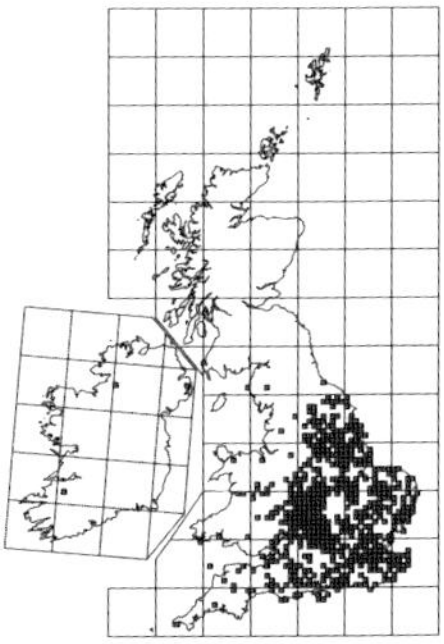

Map 40 Large flax flea beetle *Aphthona euphorbiae*.

Aphthona euphorbiae

Large flax flea beetle. A Palaearctic species, also in North Africa. In Britain, widespread and common in England, scattered in Wales and Ireland. Found on many different plant species in a range of habitats. 1.8–2.0 mm.

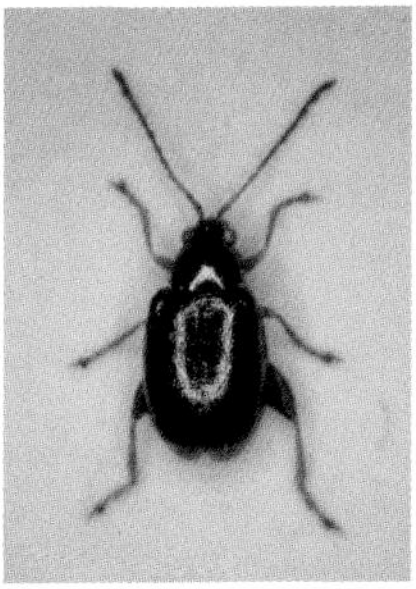

Map 41 *Longitarsus absynthii.*

Longitarsus absynthii

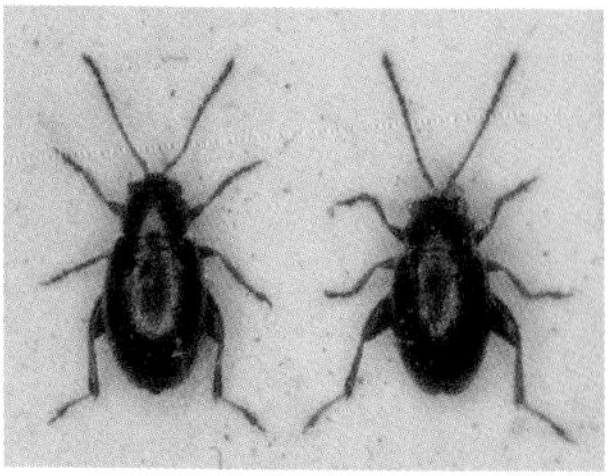

A Palaearctic species. In Britain, found in a small number of coastal locations in south-east England. Adults are found on sea wormwood *Seriphidium maritimum* and mugworts (*Artemisia* species), and occasionally yarrow *Achillea millefolium* and tansy *Tanacetum vulgare*. The larvae are undescribed but are likely to be root-feeders on the same plant species. 1.4–1.8 mm.

Map 42 Mint flea beetle *Longitarsus ferrugineus.*

Longitarsus ferrugineus

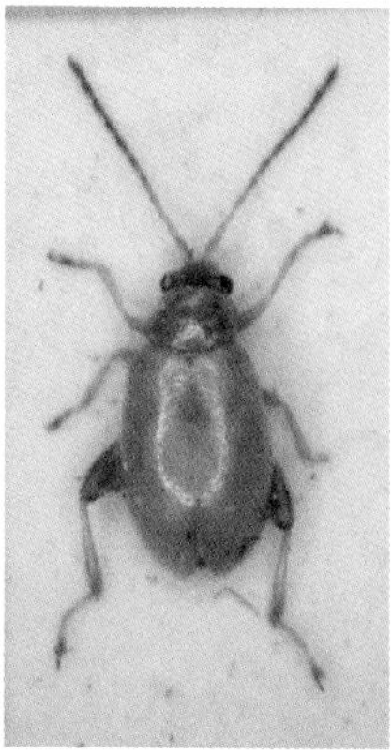

Mint flea beetle. A western Palaearctic species, also North Africa and introduced to North America. In Britain, previously widespread but localised in southern and eastern England, now sparsely scattered in a few locations in the south-east with recent records from Grays (Essex) and RAF Mildenhall (Suffolk) in the 1990s. It inhabits various, mostly damp, habitats, usually on mints, sometimes on gypsyworts (*Lycopus* species) and germanders (*Teucrium* species). Adults feed on leaves, larvae on the roots of mints only. 1.7–2.4 mm.

Map 43 *Longitarsus longiseta.*

Longitarsus longiseta

A Palaearctic species. In Britain, rare in south-east England with very few verified sites. There were several records in the early 1990s but subsequent habitat degradation may have occurred in at least one of the sites. It may now be restricted to a single site in Sussex, but could also be under-recorded as it can be difficult to identify. Therefore, targeted survey work is required. It is found on speedwells (with a possible preference for heath speedwell *Veronica officinalis*) in woodland clearings, shady grassland and fallow fields, especially bordering woodland. Adults feed on leaves, larvae probably developing at the roots, however its biology and ecology are poorly understood. 1.6–2.0 mm.

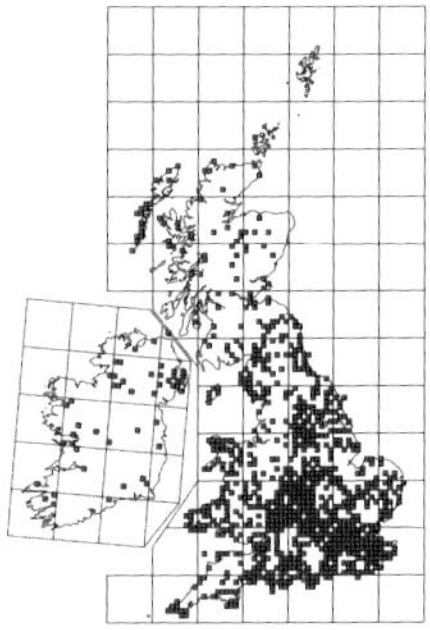

Map 44 *Longitarsus luridus.*

Longitarsus luridus

A Palaearctic species, also North Africa and introduced to North America. In Britain, widespread and common on many plant species in most habitats. 1.5–2.2 mm.

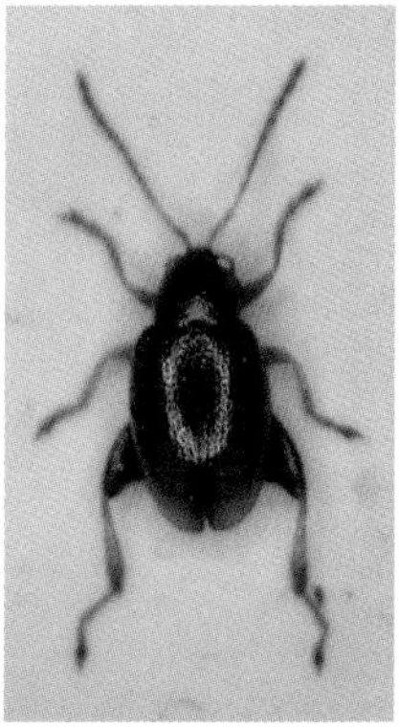

Map 45 *Longitarsus nigerrimus.*

Longitarsus nigerrimus

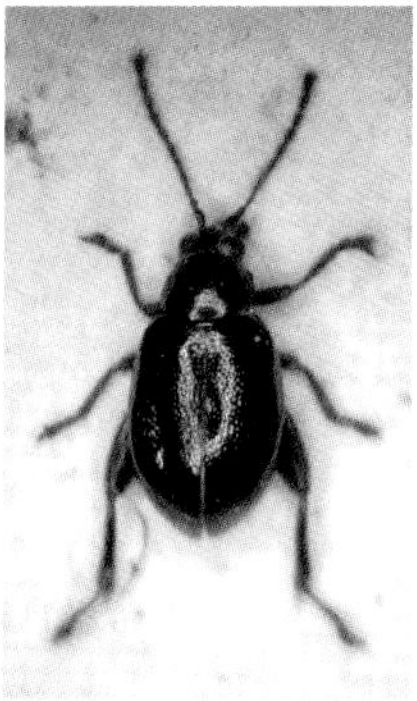

A Palaearctic species. In Britain, restricted to a few locations in the New Forest and east Dorset where it is found in boggy habitats with bladderworts (*Utricularia* species). It may also use purple moor-grass *Molinia caerulea* and cottongrasses (*Eriophorum* species) by boggy pools, but this requires confirmation. Adults feed above the water on leaves and stems, especially on lesser bladderwort *Utricularia minor*. Larvae feed wholly or partly submerged, sometimes with the rear of the abdomen exposed to the air. 1.5–2.3 mm.

Map 46 *Longitarsus obliteratoides.*

Longitarsus obliteratoides

A south-west Palaearctic species first described in 1973. In Britain, the first record was in 1992 in Pembrokeshire and it is now known from a few coastal sites in Wales and south-west England. It is known from sea cliffs, calcareous grassland and sandy beaches, with adults on thyme *Thymus serpyllum* and larvae probably on the roots, although the larva has yet to be described. 1.1–1.5 mm.

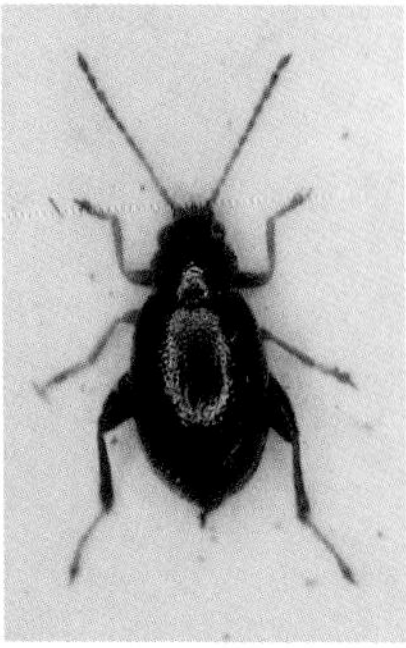

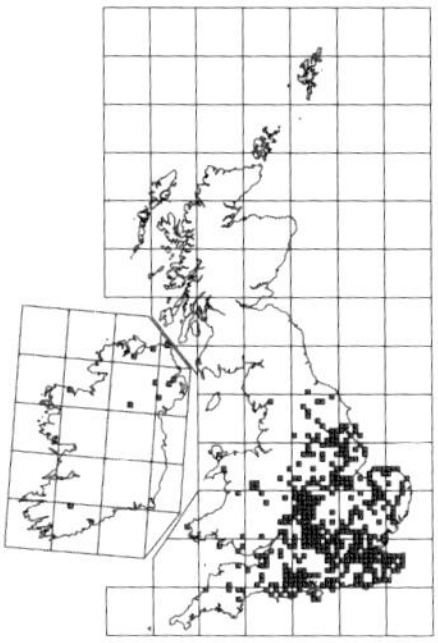

Map 47 Flax flea beetle *Longitarsus parvulus.*

Longitarsus parvulus

Flax flea beetle. A Palaearctic species, also North Africa and associated North Atlantic islands. In Britain, common and widespread in England following an expansion in range since the 1990s, probably due to increased linseed cultivation, though still rare in Wales and with only old records in Ireland. 1.3–1.5 mm.

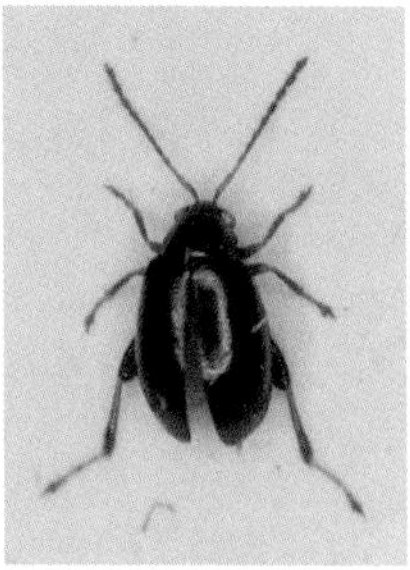

Map 48 *Longitarsus plantagomaritimus.*

Longitarsus plantagomaritimus

A north-west Palaearctic species. In Britain, scattered around the coast of England, Scotland and Wales in habitats such as saltmarshes, estuaries and dunes. Feeds on sea plantain *Plantago maritima*. 1.4–1.5 mm.

Map 49 *Altica carinthiaca.*

Altica carinthiaca

A Palaearctic species. In Britain, found in southern England on meadow vetchling *Lathyrus pratensis* in various habitats. It is possible that other plants may be used, but this is uncertain. 3.0–4.0 mm.

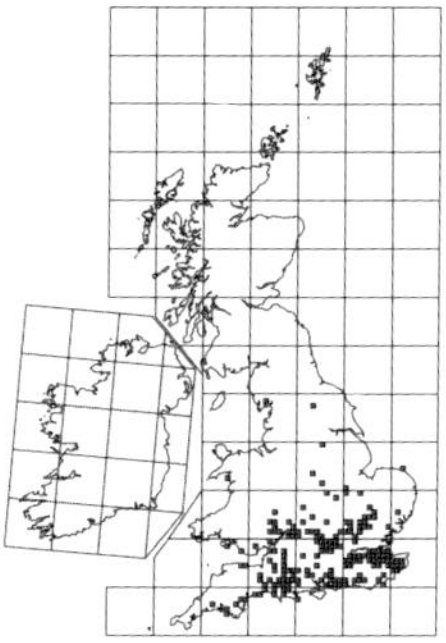

Map 50 Dog's-mercury flea beetle *Hermaeophaga mercurialis.*

Hermaeophaga mercurialis

Dog's-mercury flea beetle. A western Palaearctic species. In Britain, widespread and common in southern England, scattered in Wales with a few records further north. Found on dog's-mercury *Mercurialis perennis* in various habitats. 2.3–3.0 mm.

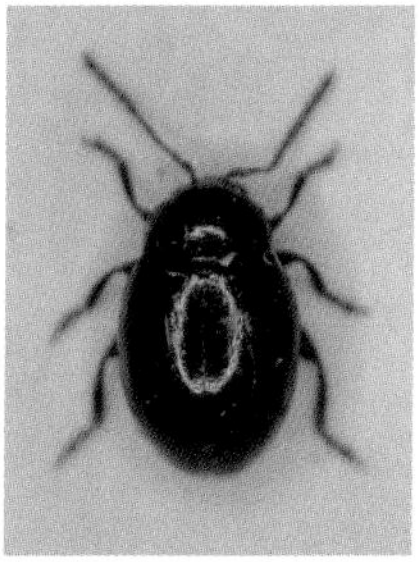

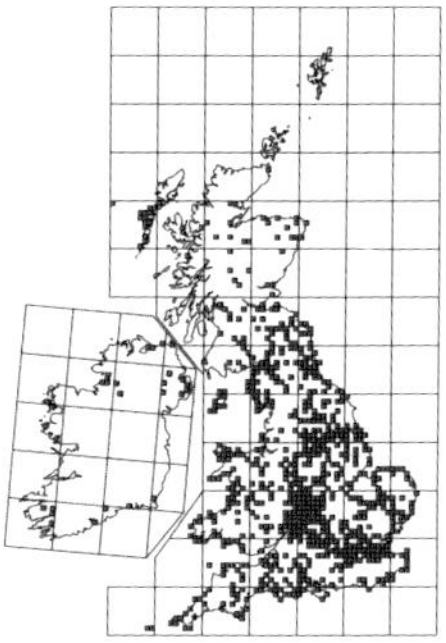

Map 51 Wheat flea beetle *Neocrepidodera ferruginea.*

Neocrepidodera ferruginea

Wheat flea beetle. A western Palaearctic species. In Britain, widespread and common on grasses including cereals, and many other plants in a range of habitats. 3.0–3.6 mm.

Map 52 *Neocrepidodera impressa.*

Neocrepidodera impressa

A western Palaearctic species, also North Africa. In Britain, scattered around the coasts of England and Wales, mainly on saltmarshes and dunes. Adults are usually found on sea-lavender *Limonium vulgare*, and occasionally on thistles. The larvae have not been described but probably develop in sea-lavender roots or stems. 4.0–5.5 mm.

Map 53 Mallow flea beetle *Podagrica fuscipes.*

Podagrica fuscipes

Mallow flea beetle. A south-west Palaearctic species, also known from North Africa. In Britain, sparsely scattered with a cluster of records around the Thames estuary. Found on mallows and hollyhocks in various habitats. 3.0–6.0 mm.

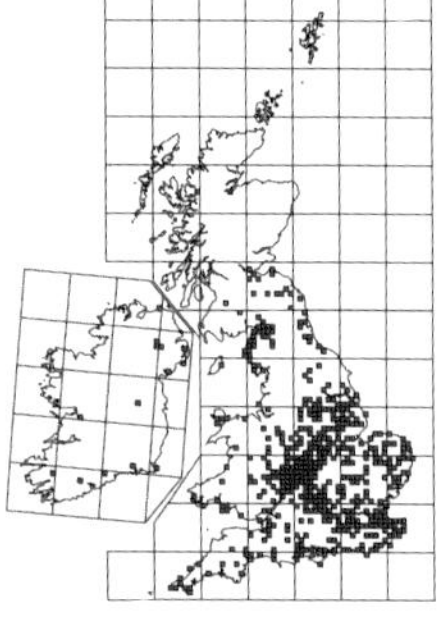

Map 54 Mangold flea beetle *Chaetocnema concinna.*

Chaetocnema concinna

Mangold flea beetle. A Palaearctic species, also North Africa and the Middle East, and introduced to North America. Common and widespread in most habitats on a wide range of plants, especially docks and other members of the Polygonaceae. It has previously been confused with *Chaetocnema picipes* and can be found together with it, though *C. picipes* is much less common. 1.8–2.4 mm.

Map 55 *Chaetocnema subcoerulea.*

Chaetocnema subcoerulea

A western Palaearctic species. In Britain, mostly found in the southernmost counties of England, on sedges and rushes in a variety of wet habitats. 1.8–2.2 mm.

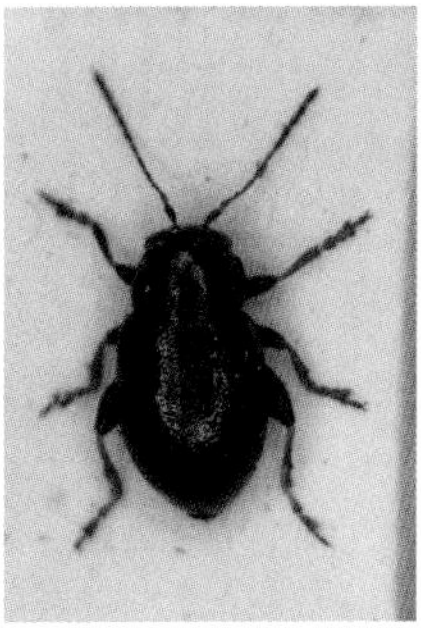

Map 56 Moss flea beetle *Mniophila muscorum.*

Mniophila muscorum

Moss flea beetle. A western Palaearctic species. In Britain, widespread but very scattered among mosses such as *Rhytidiadelphus* in woodlands, parkland and moors. The small size (maximum 1.5 mm long) and tendency to be hidden among moss means this species is probably under-recorded and further observations would help to clarify its distribution. 1.0–1.5 mm.

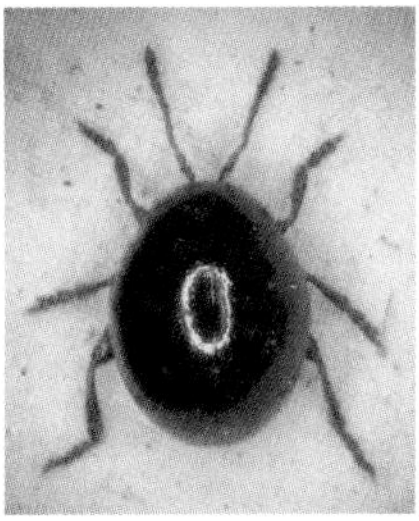

Map 57 *Dibolia cynoglossi.*

Dibolia cynoglossi

A western Palaearctic species, also North Africa. In Britain, previously known from a few sites in southern and eastern England, but this has now declined to two with recent records. These are both SSSIs on shingle – Rye Harbour (East Sussex) and Dungeness (Kent), although older records include woodland rides, clearings and margins, and chalk hillsides. Adults feed on leaves of various Lamiaceae especially hemp-nettles (*Galeopsis* species), and larvae are leaf-miners. 2.4–3.0 mm.

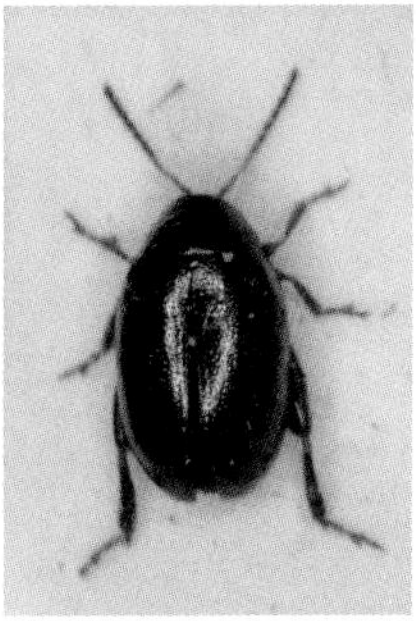

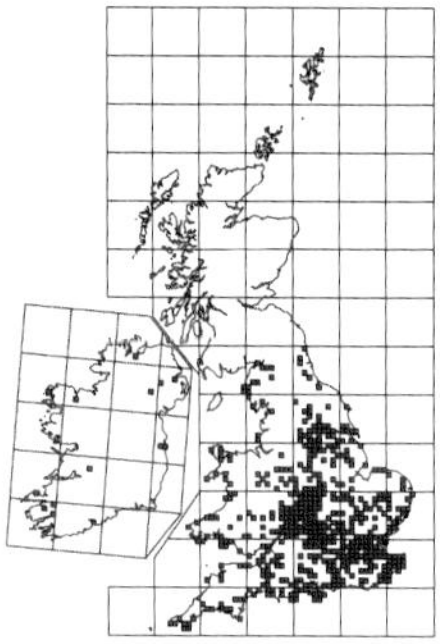

Map 58 Potato flea beetle *Psylliodes affinis*.

Psylliodes affinis

Potato flea beetle. A Palaearctic species, also North Africa and introduced to North America. In Britain, common and widespread in England and Wales, scattered in Ireland. Found in many habitats on a range of Solanaceae, especially bittersweet *Solanum dulcamara*. 2.2–2.9 mm.

Map 59 Hop flea beetle *Psylliodes attenuata*.

Psylliodes attenuata

Hop flea beetle. A Palaearctic species. In Britain, previously widespread but following a large decline at least partly due to reduced hop cultivation, now known from a small number of widely scattered locations, mainly in Kent, also Warwickshire and Nottinghamshire. It is found on and around cultivated land, especially hop-fields and margins, also woodland. Adults feed on the leaves, flowers and cones of Cannabaceae, especially hops, while early instar larvae mine roots and later instars feed on the outer surface of roots. 2.0–2.8 mm.

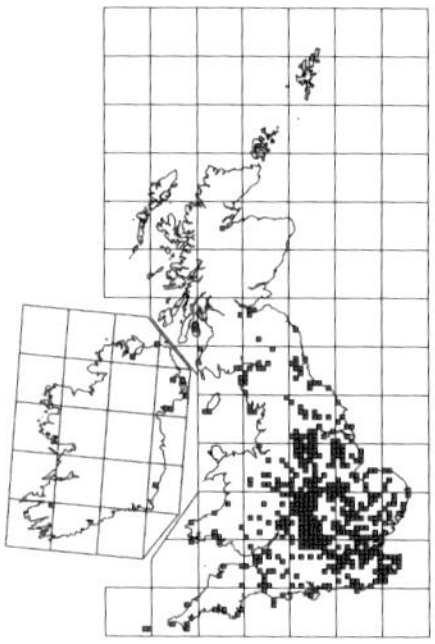

Map 60 Cabbage stem flea beetle *Psylliodes chrysocephala*.

Psylliodes chrysocephala

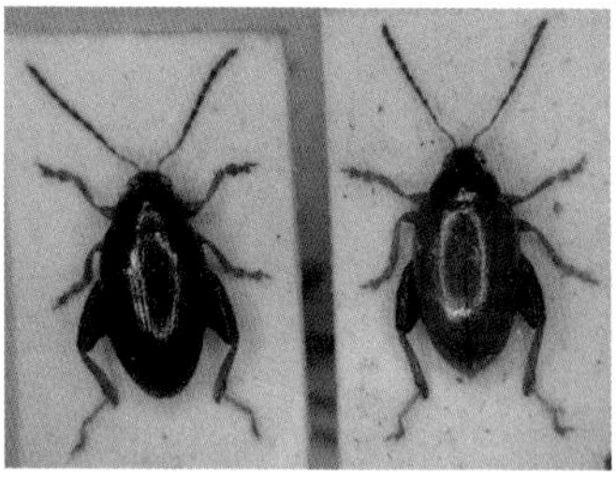

Cabbage stem flea beetle. A western Palaearctic species, also the Middle East, North Africa, and associated North Atlantic islands. Introduced to North America. In Britain, widespread and common in most habitats, especially in England on a wide variety of Brassicaceae including crops. 3.0–4.5 mm.

Map 61 *Psylliodes cucullata.*

Psylliodes cucullata

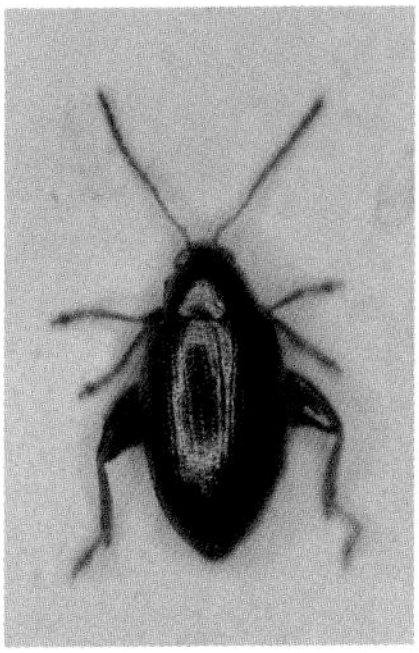

A Palaearctic species, possibly also the Middle East and North Africa, introduced to North America. In Britain, first recorded in 1991 and known from a small number of locations in south Wales. It has been found in woodland and arable fields. The host plant may be corn spurrey *Spergula arvensis*, although grasses including crops are a possibility, and both the larva and pupa remain undescribed. It is likely that this was an overlooked species rather than a recent introduction and targeted observations could help clarify its distribution, ecology and biology. 2.1–2.4 mm.

Map 62 Henbane flea beetle *Psylliodes hyoscyami.*

Psylliodes hyoscyami

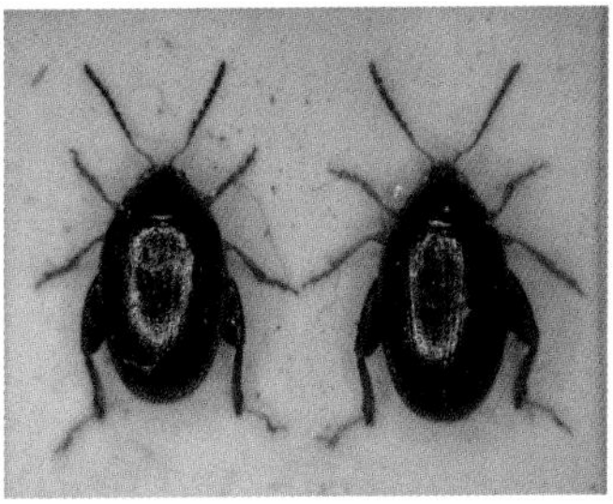

Henbane flea beetle. A Palaearctic species, also the Middle East and North Africa. In Britain, previously widespread but scattered mainly in England, but also known from Wales and Scotland. It was last recorded in 1930 in Oxfordshire following a decline along with commercial henbane crops, and is considered extinct in Britain, but is included here because recolonisation is possible as the beetle is found in France. Found in areas of disturbed ground, particularly where there is sand. It feeds mainly on henbane *Hyoscyamus niger*, occasionally on other Solanaceae such as deadly nightshade *Atropa belladonna* and bittersweet *Solanum dulcamara*. 2.9–3.5 mm.

Map 63 Lundy cabbage flea beetle *Psylliodes luridipennis*.

Psylliodes luridipennis

Lundy cabbage flea beetle. Endemic to Lundy Island off the north coast of Devon. It is found in various, mainly rocky, habitats, feeding only on Lundy cabbage *Coincya wrightii* which is also a Lundy endemic. Adults feed on the leaves, while larvae develop in petioles, midribs and stems. 3.1–3.7 mm.

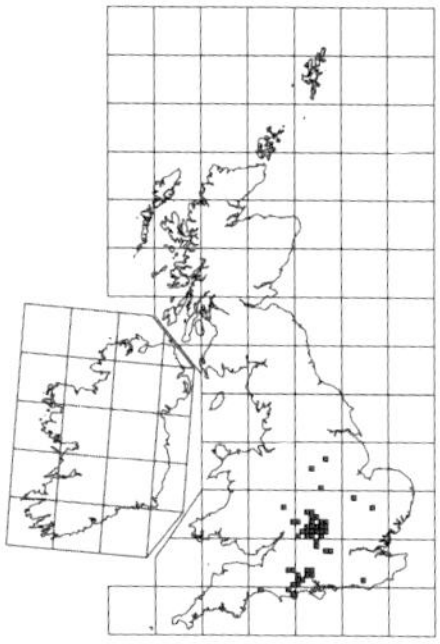

Map 64 *Psylliodes luteola.*

Psylliodes luteola

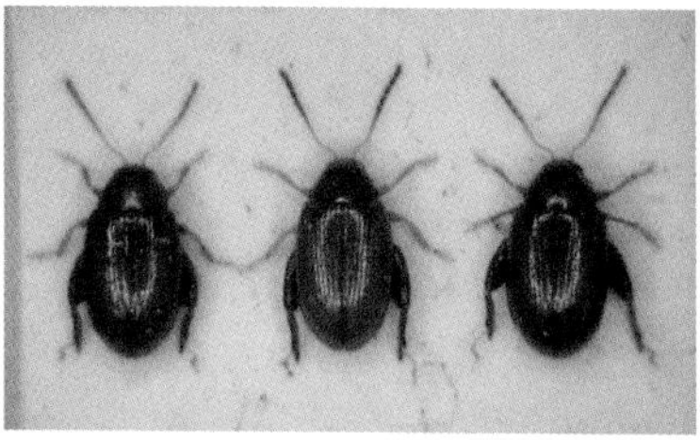

A western Palaearctic species, also the Middle East and North Africa. In Britain, mainly in central southern England in a range of habitats on wild and cultivated grasses, sometimes on trees and shrubs. It was first recorded in Oxfordshire in 1912 and most records are from this area. Few further records exist until the 1980s and it is possible that its range expanded after this time, though targeted searches may be needed to clarify its current status. 2.2–3.1 mm.

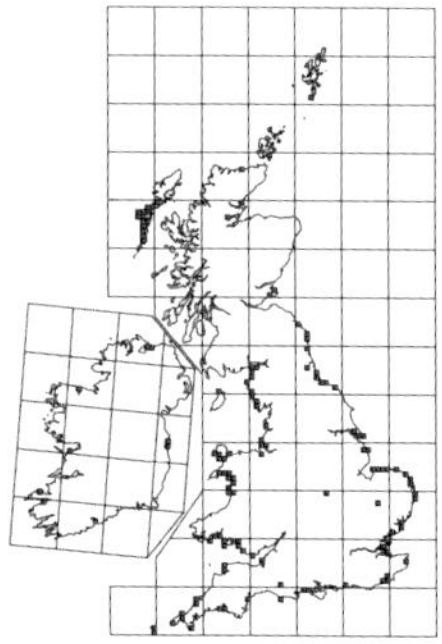

Map 65 *Psylliodes marcida.*

Psylliodes marcida

A Palaearctic species, also the Middle East and North Africa. In Britain, widespread around the coast on dunes and road verges, feeding on sea rocket *Cakile maritima* and sometimes found on other Brassicaceae. 3.2–3.8 mm.

Psylliodes sophiae

Map 66 Flixweed flea beetle *Psylliodes sophiae*.

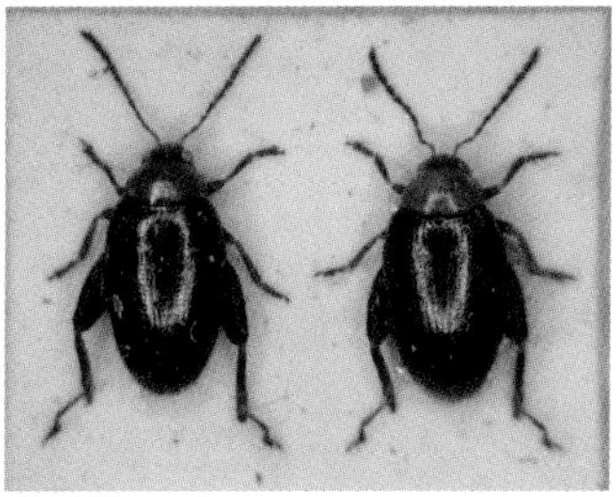

Flixweed flea beetle. A Palaearctic species. In Britain, a localised Breckland species now declined to a small number of locations in East Anglia with recent records only from west Suffolk and west Norfolk. It is found in various habitats on flixweed *Descurainia sophia*, and possibly also feeds on woad *Isatis tinctoria*. Many specimens, including any outside East Anglia, were incorrectly identified *Psylliodes chrysocephala*. 2.8–3.7 mm.

Cassida murraea

Map 67 Fleabane tortoise beetle *Cassida murraea*.

Fleabane tortoise beetle. A Palaearctic species. In Britain, now restricted to the south and west of England and Wales, although some old records reach to East Anglia. It is found in various damp habitats, including coastal ones, and is known on common fleabane *Pulicaria dysenterica* and marsh thistle *Cirsium palustre*. 6.5–9.0 mm.

Cassida sanguinosa

Map 68 *Cassida sanguinosa*.

A Palaearctic species, also North Africa. In Britain, very sparsely scattered in the southernmost counties of England. It is found in various habitats, usually near water, and sometimes on farmland. Adults and larvae feed on several species of Asteraceae, such as tansy. However, the precise habitat requirements and range of foodplants are not well understood and careful observations may help to clarify these areas. Most records are since 1980 and it may be expanding its range, but this is also uncertain, requiring further records. 5.9–8.0 mm.

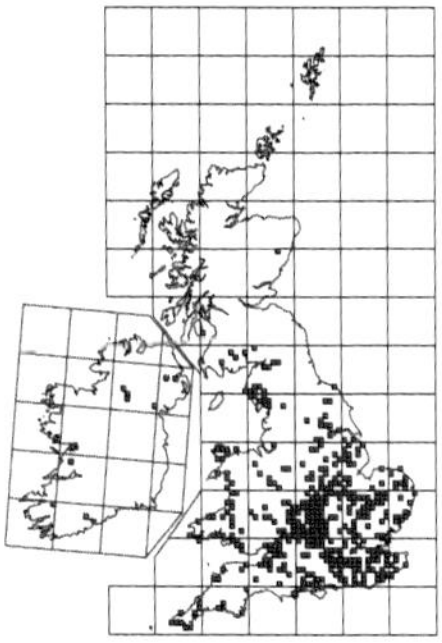

Map 69 Green tortoise beetle *Cassida viridis*.

Cassida viridis

Green tortoise beetle.* A Palaearctic species, also North Africa. In Britain, widespread, especially in England and Wales, and found in various habitats on a range of Lamiaceae, such as mints. 7.0–10.0 mm.

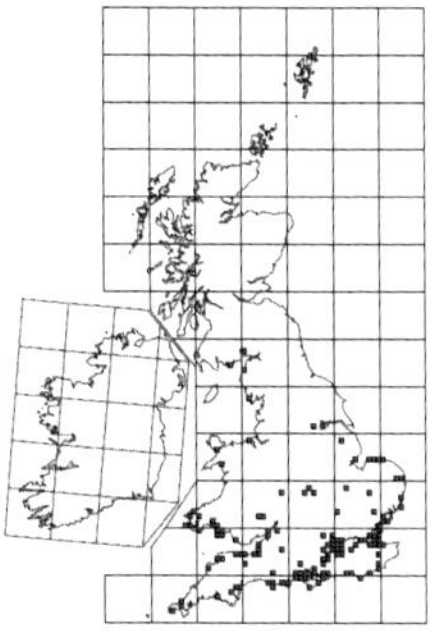

Map 70 Banded tortoise beetle *Cassida vittata*.

Cassida vittata

Banded tortoise beetle. A Palaearctic species. In Britain, scattered and patchy with most records in the south. Found in various habitats, including on the coast, on a wide range of foodplants, especially Chenopodiaceae. 5.0–6.5 mm.

* The colour of the specimen in the photo has faded from green to brown after death, which is common for many cassidines.

6 Identification of adults of British and Irish leaf beetles

6.1 Identification of families and subfamilies

The key to subfamilies is a starting point for unknown specimens of adult leaf beetles in Britain and Ireland. It is followed by keys to genera and species, and users can move straight to these if the subfamily is known. Where possible, keys use external features but in many cases, especially within the Galerucinae, species-level identification requires finer examination or dissection under a binocular microscope. A complete key to species is beyond the scope of this book, but a checklist of British and Irish species is given at the end of the chapter. For keys covering adults of all British and Irish species, see Hubble (2012). If you aren't sure whether you have a leaf beetle, check for the following features:

- 10 or 11 antennal segments
- Antennae not club-shaped
- All tarsi with four segments

Leaf beetles always have all of these features. They can sometimes be confused with ladybirds, but these have clubbed antennae and tarsi with three segments. Unwin (1984) provides an affordable key to the families of British beetles.

Sizes vary and so size ranges should be treated as guidelines. Note that sizes are given as body length, excluding the legs and antennae, but including the head. A species described as '2.5–3.5 mm' could reasonably be found as a 2.4 or 3.6 mm specimen, but not 5 mm. For colours, some coleopterists traditionally use specialist terms such as piceous (a dark colour which may have a greenish or yellowish sheen), testaceous (yellowish, usually dusky rather than bright, but applied to any yellowish to yellow-red shade), rufous (reddish), pitchy (blackish-brown but, like 'piceous', a rather variable term) and fuscous (brown to tawny-brown). I generally avoid these terms and have given colours their everyday names. The keys only cover adults. For identification of immature stages, see the list of sources given in Cox (2007).

coleopterist
someone who studies beetles (Coleoptera)

It is not always easy to tell males and females apart

from external features. However, in some species there are adaptations that help. Possibly the most frequently encountered of these are the dilated (broadened) tarsal segments in the males of a range of species. These contrast with the relatively narrow segments in females, for example in the hind legs of *Altica* (Fig. 6.1), although the segments affected and the nature of their modification varies between genera and species, and are adaptations that help the male grip the female while mating. In the Cryptocephalinae, the males also have elongate front legs, particularly the tibiae and tarsi which serve a similar function. This elongation is present in some other chrysomelids, but not as clearly. In *Bruchus* species, the middle legs of males have either one or two spurs on the tibia, while the females have none.

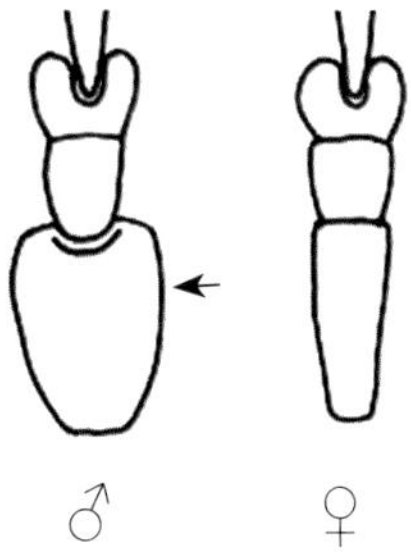

Fig. 6.1 Hind tarsi of *Altica*.

Females may show adaptations related to egg storage and laying. In *Gastrophysa* and *Galeruca*, gravid females have swollen abdomens that force the elytra apart. Females of the Cryptocephalinae and Lamprosomatinae have a deep dent in the last abdominal sternite which is where the egg is held and manipulated while being coated in the faecal material that protects it. There are further differences between the sexes of individual species that are beyond the scope of this book, for example the adaptations to male antennal segments in some *Phyllotreta* (Hubble, 2012).

The terms used for the various parts of the body are indicated on Figs. 1.1, 1.2, 1.9 and 1.10.

Key A Subfamilies and small families

1 Head covered by pronotum, not visible from above. Mouthparts directed backwards. Elytral margins flattened, forming a wide rim. In many species, the whole beetle has a flattened appearance, though some may be more convex. Often rounded in dorsal view, sometimes broadly oval, never elongate. Known as 'tortoise beetles'. 4.0–10.0 mm long (see Fig. 1.3 for the shape mentioned here) **Key B Subfamily Cassidinae** (p. 81)

– Not with flattened elytral margins or the general form indicated above. The head may or may not be covered by the pronotum **2**

2 Eyes deeply notched at the front. Antennae usually serrate (saw-like), sometimes filiform (shaped like a thin filament). Elytra generally blackish or brownish with a mottled pattern, 10 distinct striae (elongate grooves or lines), and recumbent pubescence (non-erect, fine hairs). Head usually deflexed (bent downwards), neck long with front of head extended into a short, wide, flattish rostrum or 'muzzle'. Top of head usually with a central longitudinal ridge. Hind femora often swollen or with teeth on the underside (but femoral development **not** as in the 'flea beetles' of subfamily Galerucinae, tribe Alticini) (see Fig. 1.9 for features of the head mentioned here

Key C Subfamilies Bruchinae and Amblycerinae (p. 81)

Confirmatory characters: 1.7–5.3 mm long (with head deflexed). Four tarsal segments on front, middle and hind legs, with some two-lobed segments. Often associated with Rosaceae, Fabaceae and Apiaceae. Commonly known as seed beetles (sometimes other names such as 'bean weevils').

– Eyes not deeply notched at the front; may be notched at the inner edge (towards the midline between the eyes) or entirely un-notched. Head not extended into a 'muzzle'. Other features variable, but not combined as above **3**

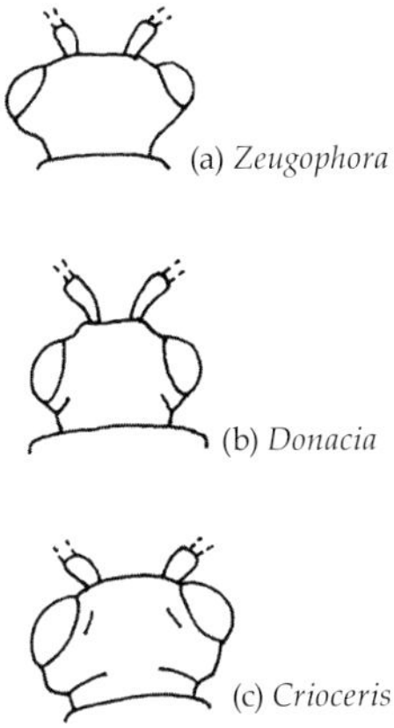

A.1 Constriction of the head

3 Head clearly (though variably) angled and constricted behind eyes (A.1). Pronotum without side margins **4**

– Head without this constriction, or if with such a constriction, 5–7 mm long and yellow with four black spots on the elytra (*Phyllobrotica quadrimaculata*). Pronotum usually with side margins (a clearly defined rim on either side), though not in *Bromius*, *Orsodacne* and some *Cryptocephalus* **6**

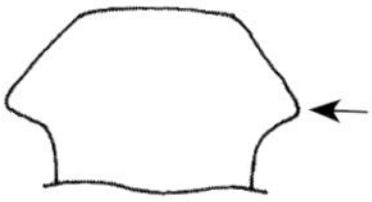

A.2 *Zeugophora* pronotal spine.

4 Elytra with punctures in rows **5**

– Elytra with random punctures

Family Megalopodidae (Subfamily Zeugophorinae)

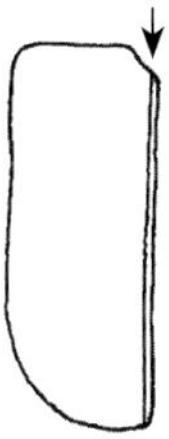

A.3 *Zeugophora* elytral suture.

Confirmatory characters: 2.5–4.0 mm long and weakly elongated. Yellow pronotum coarsely punctured with a broad central lateral spine on each side (A.2). Elytra irregularly and coarsely punctured but not striate. Elytral suture has a rim along the entire length (A.3). Claws with a blunt appendage beneath (A.4) though this may be hard to see. Other features broadly similar to Orsodacnidae. In Britain, a single genus *Zeugophora*.

A.4 *Zeugophora* tarsal claws.

A.5 Donaciine antennal segments.

shoulders
the outer front angles of the elytra, forming bulges in some species

5 First antennal segment clearly elongate and considerably longer than segment 2 (A.5). Scutellum slightly pubescent. Eyes not notched

Key D Subfamily Donaciinae (p. 83)

Confirmatory characters: 4.5–13.0 mm long. Of characteristic appearance (see Fig. 1.5 for the overall shape), usually with metallic coloration and long antennae, they are unlikely to be confused with any other chrysomelid group. They are usually active in sunshine and may be readily visible in large numbers. Commonly known as reed beetles.

– First antennal segment rounded to oval; may be longer than segment 2, but not clearly elongate. Scutellum hairless. Eyes notched on inner edge, though this may be slight (e.g. in *Oulema*)

Key E Subfamily Criocerinae (p. 84)

Confirmatory characters: Distinctive; narrow, elongate with rectangular elytra (see Figs. 3.10, 3.12 for the overall shape) and often vividly coloured (some are metallic). Pronotum narrower than elytra at shoulders and never bordered. Includes lily and asparagus beetles.

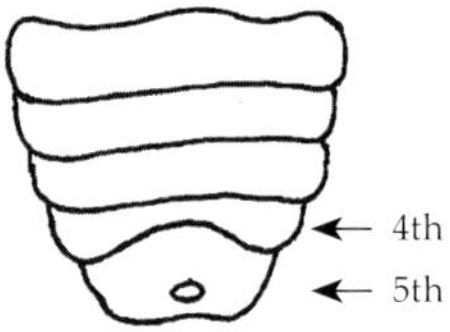

A.6 Cryptocephaline sternites 4 and 5.

6 Fourth abdominal sternite strongly constricted in the middle (A.6). If only slightly constricted, then antennae serrate. Females with a round dent in the 5th abdominal sternite

Key F Subfamily Cryptocephalinae (p. 85)

Confirmatory characters: Parallel-sided with the head largely or entirely hidden from above by the bulging pronotum, especially in *Cryptocephalus*. Antennae long and filiform in *Cryptocephalus* (most species, and the majority of specimens), serrate in *Labidostomis*, *Clytra* and *Smaragdina* (tribe Clytrini). All are distinctly coloured and mostly stenophagous (feeding on a single food-plant or a limited range).

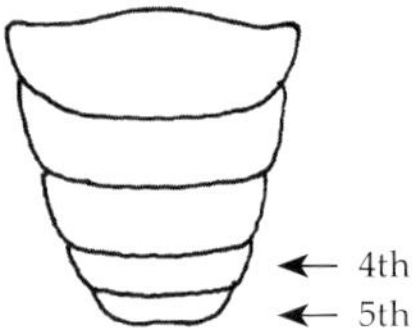

A.7 Abdominal sternites 4 and 5.

– Fourth abdominal sternite not, or only slightly (clearly less than in A.7), constricted in the middle. Antennae not necessarily serrate. Females without a round dent in the centre of the 5th abdominal sternite (A.7) 7

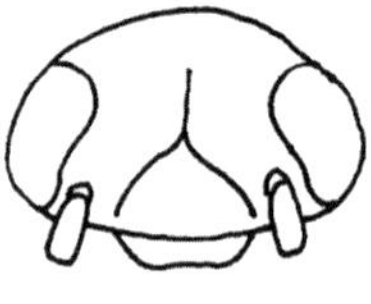

A.8 Antennal bases widely separated.

7 Antennal bases beneath side margin of head, separated by about twice the length of the first antennal segment (A.8) 8

A.9 Antennal bases close together.

– Antennal bases on forehead (i.e. between or in front of the eyes), no further apart than the length of the first antennal segment (A.9) . Includes the flea beetles with enlarged hind femora and the ability to jump
Key G Subfamily Galerucinae (p. 86)

8 Front coxae rounded (A.10) **9**

– Front coxae transverse (A.11) **10**

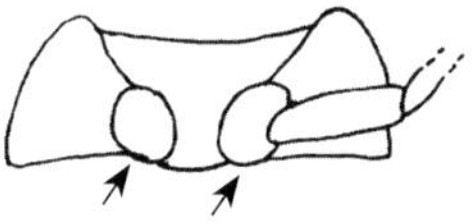

A.10 Rounded coxae.

9 Antennal segments 7 and 9–11 widened (A.12). Elytral epipleura with three depressions, hind one deepest (A.13) **Subfamily Lamprosomatinae**

Confirmatory characters: 2.5–3.5 mm, oval, convex and hairless. Brassy, shiny black or brilliant bronze. Antennae short, black and bluntly-toothed; second segment red-brown, segments 7 and 9–11 swollen. Pronotum and elytra finely punctured. In ventral view, elytral epipleura with three depressions, the hind one deepest. A single species *Oomorphus concolor*. Egg-shaped oval (A.14), convex and hairless. Usually brassy. Antennae short, black and bluntly-toothed; first segment red-brown. Pronotum and elytra finely punctured. Tibiae dilated and smooth on outer side (A.15). Males and females difficult to separate. Widespread across southern England and Wales, some scattered northerly records. Adults may be beaten from ivy growing on tree trunks.

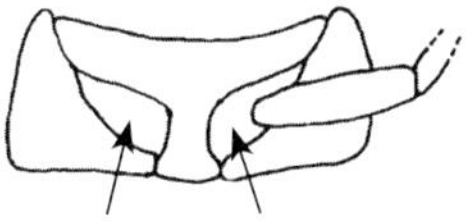

A.11 Transverse coxae.

A.12 *Oomorphus* antenna.

– Antennal segments enlarge more or less gradually towards the tip. No epipleural depressions
Subfamily Eumolpinae

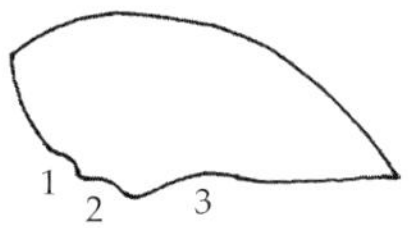

A.13 *Oomorpus* epipleura.

Confirmatory characters: 5.0–6.0 mm, elongate (for chrysomelids), dark-coloured, non-metallic with pale pubescence (fine downiness). Elytral epipleura without depressions, antennae with segments gradually increasing in size (both length and width, but most clearly in length) towards the tip. A single species, *Bromius obscurus*. Uniformly black with dull yellow-grey pubescence. Four basal antennal segments orange-red. Males and females difficult to separate. Adults feed on various willowherbs, making 'scribbling' marks. Endangered (RDB1), found in a single 10 km square on the Cheshire–Staffordshire border.

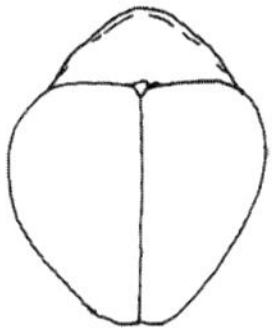

A.14 *Oomorphus* dorsal view.

epipleura
the parts of the side edges of the elytra that wrap round to the underside of the beetle (singular: epipleuron)

A.15 *Oomorphus* tibia.

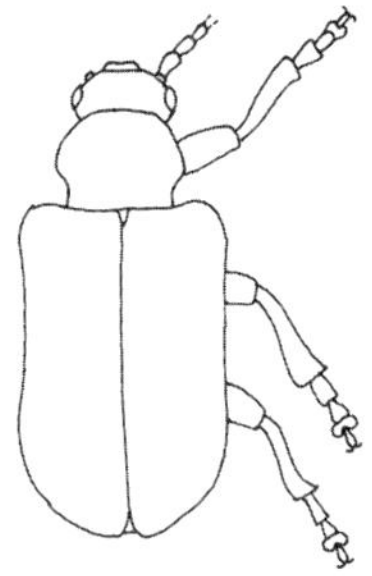

A.16 *Orsodacne* pronotum.

10 Head slightly elongate, forming a short 'muzzle' in front of antennal bases. Pronotum without side margins, narrowed towards the rear, considerably narrower than front of elytra (A.16)

Family Orsodacnidae (Subfamily Orsodacninae)

Confirmatory characters: 4.0–8.0 mm long. Pale brown to black. Head with a narrow neck so that the eyes are some distance from the front edge of the pronotum. Pronotum without lateral teeth. Elytra elongate and randomly punctured, rounded at apex (in some specimens elytral punctures tend to run in longitudinal lines near the middle but overall the randomness should be clear). Antennae filiform with segment 1 bulbous, 2 squarish (quadrate) and 3–11 elongate; segments 7–10 about 1.5 times longer than wide (A.17). Antennae inserted behind the mandibles but forward of the front edge of the eyes; insertions widely separated, a little wider than the length of the basal segment. Eyes entire (not notched). Front coxae transverse (visible from the side). Third tarsal segment bilobed, on middle and hind legs deeply so (A.18). Colour varies from dark blue to yellow or a combination of these and a casual glance in the field suggests a robust cantharid (soldier beetle), but broader and more convex. In Britain, a single genus *Orsodacne*.

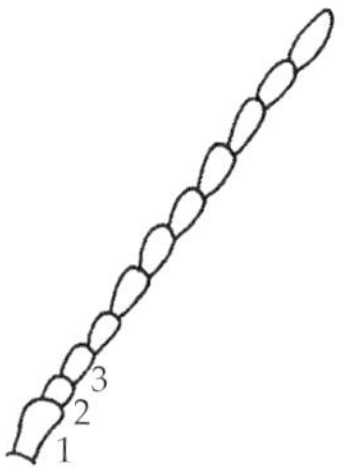

A.17 *Orsodacne* antenna.

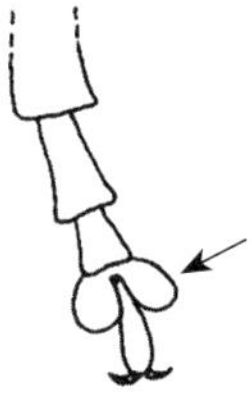

A.18 *Orsodacne* tarsus.

– Head not elongate in front of antennal bases. Pronotum with side margins, not constricted at rear, where it is, at most, slightly narrower than front of elytra (A.19)

Key H Subfamily Chrysomelinae (p. 95)

Confirmatory characters: 2.5–18.0 mm. Brightly coloured and usually shining metallic species. Hairless, generally round or oval (some may be unusually elongate for chrysomelids, but still oval) and very convex – often the 'typical' leaf-beetle shape and colour. Antennae often gradually thicken towards the tip; inserted beneath side margin of head and widely placed, further apart than the length of the basal joint. Head not elongate in front of antennal bases. Pronotum with a side margin, not narrowed at the rear, and no more than slightly narrower than the front of the elytra. Mostly stenophagous (feeding on a single food-plant or a limited range), found by beating shrubs or trees or general sweeping.

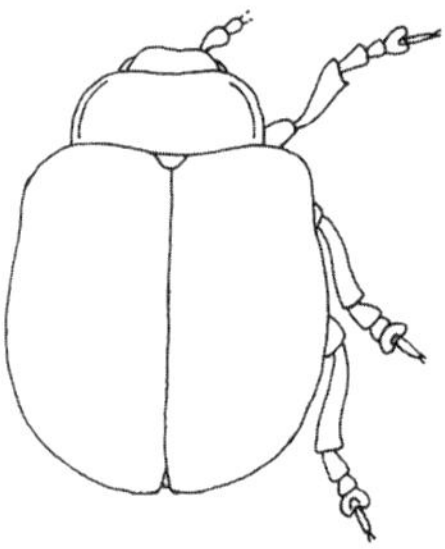

A.19 Chrysomeline pronotum.

6.2 Keys to genera

Keys lead to species where these are the only British members of the genus.

Key B Subfamily Cassidinae

Males and females are difficult to separate but identification is possible using surface characters, so dissection is not required.

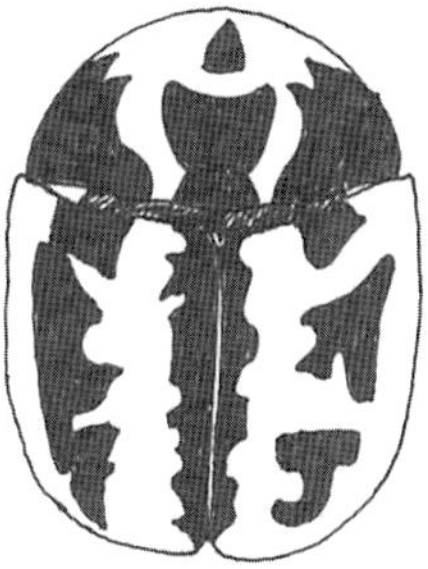

B.1 *Pilemostoma* pronotum.

1 Elytra red with irregular black stripes and suture; sometimes black with a few red spots, or red with a few small black spots. Newly emerged specimens sometimes greenish instead of red. Punctures in irregular lines. Pronotum no more than slightly narrower than elytra; shiny with black spots, and edges strongly raised (B.1). Legs black. Deep antennal groove running alongside head on underside of pronotum. 4.5–6.5 mm **_Pilemostoma fastuosa_**

Scarce in southern England and south Wales on Asteraceae, especially ploughman's-spikenard *Inula conyzae* and common fleabane *Pulicaria dysenterica*, sometimes common ragwort *Senecio jacobaea*, possibly also on mints *Mentha*, in various habitats.

– Colour and pattern not as above **_Cassida_ species**

Key C Subfamilies Bruchinae and Amblycerinae

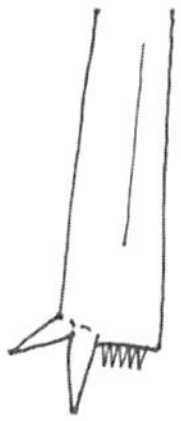

C.1 Amblycerine tibial spurs.

1 Hind tibiae with two long movable apical spurs (C.1). Hind coxae about twice as wide as hind femora, which are narrow without ventral teeth **Subfamily Amblycerinae**

Confirmatory characters: In Britain, a single species, *Zabrotes subfasciatus*, the Mexican bean beetle). Males with uniform pale brown pubescence (fine downiness) on the dark cuticle; females with pronotum and elytra clearly marked with a white pubescent pattern on the dark cuticle. Elytra squat and broad. Hind femora with a longitudinal ventral groove bordered by sharp, untoothed keels. 1.8–2.5 mm. Originally from the Americas, it is now found in many tropical and subtropical areas and is a pest of beans (*Phaseolus*). Although recorded on imported produce, it is an introduced species which is not considered established and so is not listed in Duff (2008), although it is described and figured in Cox (2001) and included, without a map, in Cox (2007).

C.2 Bruchine tibial spur.

– Hind tibiae with a single long apical spur (C.2), or two short apical spurs (C.3). In either case the spurs are immobile (in live and dead specimens). Hind coxae less than twice as wide as hind femora, which are thickened (with or without one or more teeth)
2 (Subfamily Bruchinae)

C.3 Bruchine tibial spurs.

2 Pronotum wide, sides at least slightly notched, sometimes with a small tooth at the front edge of the notch (C.4). Hind femora ventrally with a large tooth on the inner keel. Males usually with mid tibiae usually bearing a spur, or 1 or 2 teeth, at and/or near the apex
***Bruchus* species**

– Pronotum not notched; narrows conically towards the front (C.5), less so in *Bruchidius villosus*, with all dark legs and uniform pubescence dorsally. Hind femora not as above. Males without teeth or a spur on the mid tibiae **3**

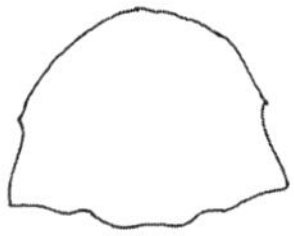

C.4 *Bruchus* pronotum.

3 Rear of pronotum with a double callus (bump) in the centre bearing dense white pubescence. First hind tarsal segment clearly longer than others combined. This difference in length is variable but should be fairly easy to determine, so if you cannot tell whether the segment is longer than the others combined, follow the second half of this couplet. Hind femora ventrally usually with a tooth on both inner and outer keels
***Callosobruchus* species**

– Rear of pronotum not as above. First hind tarsal segment may be equal to, shorter than or no more than slightly longer than others combined **4**

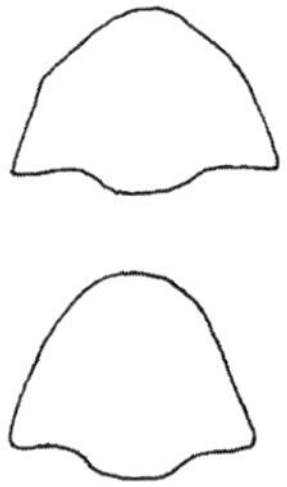

C.5 Bruchine pronotum (not *Bruchus*).

4 Hind femora ventrally with large backwards-pointing tooth in the apical half, followed by two or three (rarely one) smaller teeth (C.6). Pronotum with scale-like pubescence obscuring otherwise clear microsculpturation (textured patterning at a very small scale, smaller than any main punctures)
Acanthoscelides obtectus (dried bean beetle)

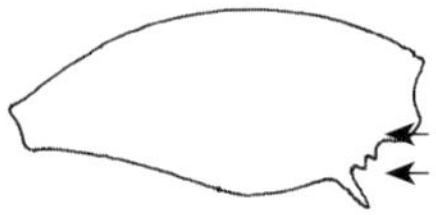

C.6 *Acanthoscelides* femur.

Confirmatory characters: Antennae dark grey, segments 1–5 and 11 often reddish. Elytral pubescence forming numerous faint spots. Most of abdomen, tips of elytra, and legs yellowish-red, except underside of mid and hind femora (black). 2.5–3.0 mm. Since the 1920s, an introduced pest of dried foods. Scattered, mostly in south and south-east England.

– Hind femora without teeth ventrally (rarely with a very small spine). Pronotum without scale-like pubescence; microsculpturation not obscured ***Bruchidius* species**

Key D Subfamily Donaciinae

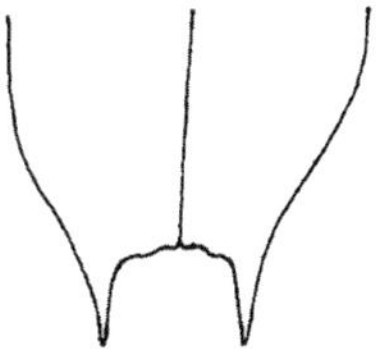

D.1 *Macroplea* elytral apex.

Males have front tarsi dilated; this is not the case in females. See individual species accounts for other features such as femoral teeth which may be used to separate males and females.

1 Dull yellowish (non-metallic) with grey or black head and elytral striae (elongate grooves or lines). Elytral apex sinuate (wavy) and toothed at the rear corners (D.1). Tarsi long and slender with 3rd segment simple (not lobed). Final tarsal segment (with claws) longer than others combined (D.2). 5.0–7.5 mm ***Macroplea* species**

Distribution: Usually under water on plants.

D.2 *Macroplea* tarsus.

– Metallic colour, elytra without tooth at the end. Tarsi not elongated; 3rd segment lobed. 5.0–12.0 mm 2

Distribution: Usually on water plants.

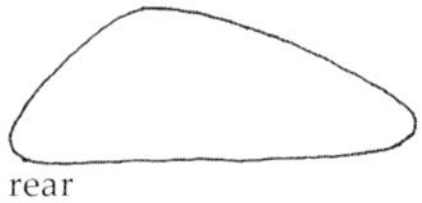

D.3 *Plateumaris* elytra.

2 Legs relatively stout, hind femora typically 2.5–3 times as long as their widest point (excluding femoral teeth), and not strongly constricted towards the base. Elytra dorsally 'vaulted' and rounded at the end; widest after the middle when viewed from side (D.3), though may be less distinctly widened than in the figure. Suture sinuate near tip (D.4), but you may need to look very closely as this feature is small and can be difficult to see. Mandibles protruding ***Plateumaris* species**

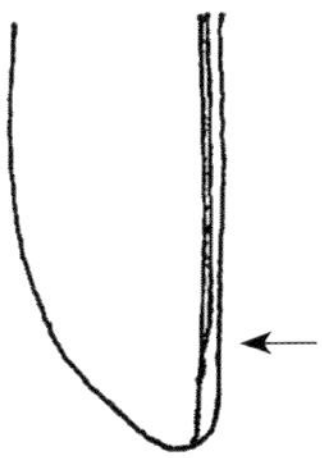

D.4 *Plateumaris* elytral suture.

– Legs relatively spindly, hind femora typically more than 3 times as long as their widest point (excluding femoral teeth), and may be strongly constricted towards the base forming a club-like shape. In some species, particularly *Donacia versicolorea* (but also *D. obscura* and *D. dentata*, and especially males), the femoral length:width ratio may overlap with that of *Plateumaris*. Elytra dorsally somewhat flattened and truncate at the end, tapering more or less evenly when viewed from the side (D.5). Suture straight to the tip. Mandibles short, not protruding ***Donacia* species**

D.5 *Donacia* elytra.

Key E Subfamily Criocerinae

For most species, separation of males and females is not straightforward without dissection, however *Crioceris asparagi* males have longer, more strongly curved front tibiae than females.

1 Head, legs and underside black, pronotum and elytra bright red. Pronotum strongly narrowed at, or just behind, the middle. 6–8 mm, the largest British criocerine **_Lilioceris lilii_** (lily beetle)

On lily and various other plants in the family Liliaceae, almost always in gardens and plant nurseries.

– Elytra not red. 3.0–6.5 mm **2**

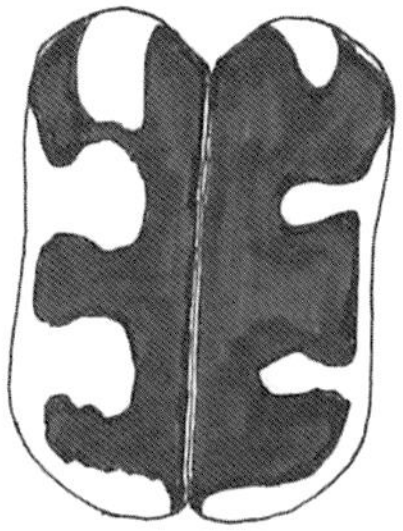

E.1 *Crioceris asparagi* pattern.

2 Head blue-black, often with a metallic green reflection. Pronotum red, often with a dark spot. Elytra predominantly blue-black, each with yellow margins and apices and three yellow marks of variable size (E.1). 5.0–6.5 mm **_Crioceris asparagi_** (asparagus beetle)

Widespread and locally common on asparagus (*Asparagus officinalis*), wild and cultivated, especially in the south-east. Note that the continental *C. duodecimpunctata* has been found in Britain but has not become established. It is orange-yellow with dark tarsi and antennae, and 12 black spots on the elytra.

– Not as above; often dark metallic blue or similar. 3.0–5.0 mm **3**

E.2 *Lema* pronotum.

3 Pronotum constricted at the mid-point (E.2) or occasionally further back but in any case the constriction of the sides does not line up with the transverse furrow. 3.5–5.0 mm **_Lema cyanella_**

Distribution: Widespread on various thistles (*Cirsium, Carduus,* and the introduced *Silybum*), in a wide range of habitats.

– Pronotum not constricted at the mid-point; narrowed further back and this constriction lines up with the transverse furrow (E.3). 3.0–4.8 mm **_Oulema_ species**

E.3 *Oulema* pronotum.

Key F Subfamily Cryptocephalinae

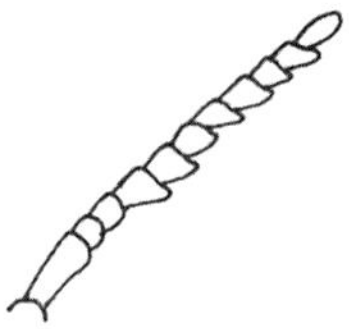

F.1 Clytrini antenna.

1 Antennae long, filiform. 2.0–8.0 mm *Cryptocephalus* **species**

– Antennae serrate (saw-like) from the 3rd or 4th segment to the tip (F.1). 2.5–9.5 mm **2** (Tribe Clytrini)

2 Elytra blue with fine, dense puncturation. Females have a deep circular pit located centrally in the apical abdominal sternite (underside plate); this is absent in males. 2.5–4.0 mm ***Smaragdina affinis***

Confirmatory characters: Blue-black head. Pronotum black or blue-black with broad orange to red-brown side margins. Endangered (RDB1). On hazel (*Corylus*), sometimes birch (*Betula*) and Asteraceae in broadleaved woodland and marshy thickets near rivers; known only from a few sites in Oxfordshire and Gloucestershire; no records since 1965. Note that *Smaragdina salicina* is known from a single specimen collected in Buckinghamshire in 2010 by sweeping in mixed deciduous hedgerow and scrub habitat on a SW-facing chalk grassland slope (Hubble & Murray, 2011). It is easily separated from *S. affinis* by having an entirely orange to red-brown pronotum. It is also larger at 5.5 mm (for the single British specimen) and differs in the structure of the aedeagus, which is required in order to separate it from some other non-British European species.

aedeagus
part of the male genitalia, a structure analogous to a penis

– Elytra yellowish-brown or yellowish-red, with or without dark spots. 6.0–9.5 mm 3

3 Head, pronotum, legs and body dark metallic-greenish or greenish-blue, elytra yellowish-brown ***Labidostomis tridentata***

Confirmatory characters: Males have larger heads and longer mandibles with a dorsal projection, the front tibiae are longer and more strongly curved than the other tibiae, and tarsal segments 1 and 2 are longer than the same segments on other legs. Females have a deep circular pit located centrally in the apical abdominal sternite; this is absent in males. 6.0–9.0 mm. Distribution: Endangered (RDB1) and may be extinct. Rough open ground in woodland; adults usually on birch (*Betula*), feeding especially on the leaves of 5-year old saplings. Known only from a few scattered sites in Hampshire, Kent, Sussex, Worcestershire and Yorkshire.

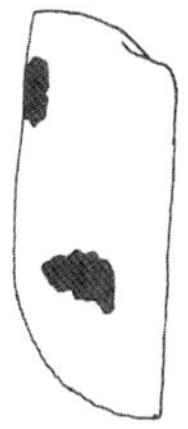

F.2 *Clytra quadripunctata* pattern.

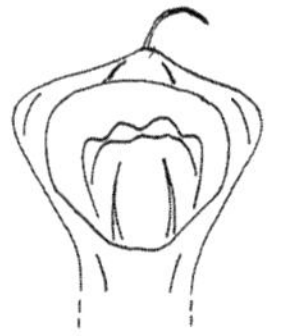

F.3 *Clytra quadripunctata* aedeagus.

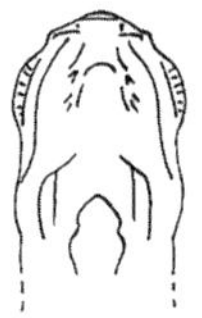

F.4 *Clytra laeviuscula* aedeagus.

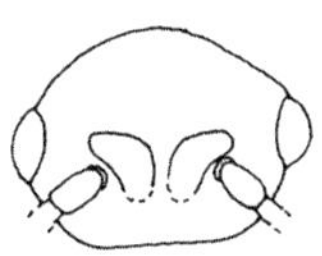

G.1 Galerucini antennal bulges.

G.2 Galerucini typical dorsal view.

– Pronotum, scutellum, body and legs black, head black with a red spot behind the eyes. Elytra yellowish-red with black spots (F.2) ***Clytra quadripunctata***

Confirmatory characters: Females have a deep circular pit located centrally in the apical abdominal sternite (as for *Labidostomis tridentata*); this is absent in males, though (unlike *L. tridentata*) there may be a broad, shallow, smooth depression. Pronotum with side margins flattened, broad (especially to the rear) and clearly punctured. Aedeagus elongate with the tip clearly widened in dorsal or ventral view (F.3). 7.4–9.5 mm. Associated with various ant species; found on trees, shrubs, cock's-foot (*Dactylis glomerata*) and bracken (*Pteridium aquilinum*) in various woodland types, widespread but not common. *Clytra laeviuscula* is 7.5–11.5 mm, has the pronotum with side margins narrow and smooth, and the aedeagus short and slightly widened at the tip in dorsal or ventral view (F.4). However, it is probably extinct in Britain, the last record being from Berkshire in 1895. It is also associated with various ant species and previously known from a range of trees and shrubs in Caledonian pine and birch woodland, and on chalk grassland.

Key G Subfamily Galerucinae

In many cases, dissection of males is required for accurate identification to species. Features useful for separating males and females are given where present. However, in some cases, clear features may not be easy to see, or may be absent. In such cases, the shape of the rear abdominal segment may prove useful as this tends to be smoothly rounded in females, but often exhibits some lateral sinuosity, incision, indentation or other interruption of smoothness in males.

1 Hind femora not thickened; unable to jump. Pronotum without impression or furrow on or just in front of the rear edge. Bulges at/above bases of antennae stretch forwards and between bases (G.1), though their shape is variable **2 Tribe Galerucini** (G.2) (p. 87)

– Hind femora thickened; able to jump and so known as the 'flea beetles'. In species with hind femora only slightly thickened, the pronotum has an impression or groove on or just in front of the rear edge (G.3, G.4),

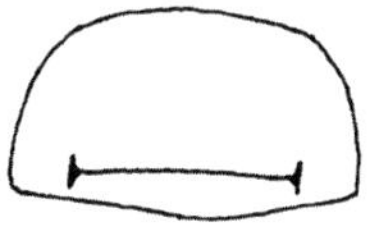

G.3 Alticini pronotal groove.

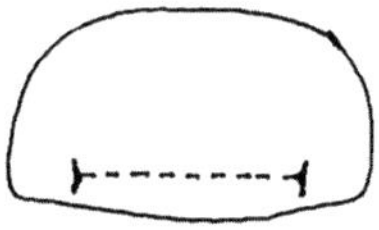

G.4 Alticini pronotal groove.

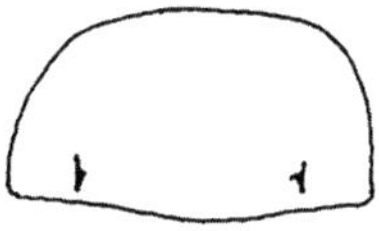

G.5 Alticini pronotal furrows.

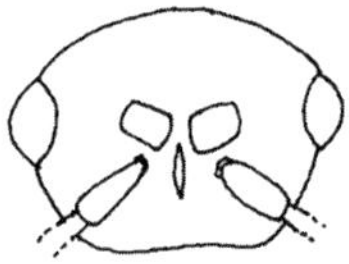

G.6 Alticini frontal keel.

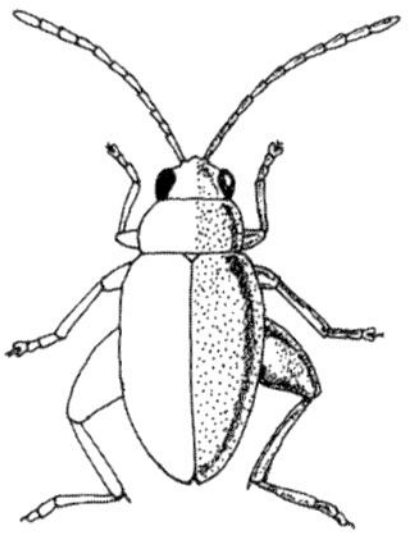

G.7 Alticini typical dorsal view.

shoulders
the outer front angles of the elytra, forming bulges in some species

sometimes only lateral longitudinal furrows (G.5). Bulges at/above bases of antennae do not stretch forwards, although they vary in shape and there may be a longitudinal keel between the bases (G.6) **12 Tribe Alticini** (G.7) (p. 90)

Tribe Galerucini

2 Elytra metallic green, blue, purple, bronze or copper. 5.0–7.0 mm **3**

– Various colours, shiny or dull, but not metallic. 3.0–10.0 mm **4**

3 Elytra metallic green or blue, rarely coppery. Head and pronotum yellow-brown (dark green to blackish spot on the top of the head). Antennae, tarsi and tips of tibiae usually dark brown to blackish; rest of legs yellow-brown, occasionally legs and antennae entirely reddish-brown. Elytra densely and moderately coarsely punctured, pronotum smooth and shiny. Pronotum approximately rectangular and twice as wide as long. 5.0–7.0 mm ***Sermylassa halensis***

Widespread on bedstraws (*Galium*), sometimes calamints (*Clinopodium*), in a wide range of habitats.

– Dorsal surface, legs and antennae deep metallic blue with violet reflection (sometimes purple or bronze), though specimens may lose their colour and appear dull black. Elytra widened towards the rear and with fairly distinct shoulders. Dorsal surface finely and densely punctured. Pronotum narrower than elytra and very short. 6.0–7.0 mm ***Agelastica alni*** (alder leaf beetle)

Distribution: Very rare, previously considered extinct but found in the Manchester area in 2004 and since in Cheshire, also more recently in Hampshire. In open sunny locations in wetlands, especially alder carr, also river banks and wet woodland flushes. On young alder, sometimes hazel, occasionally other small trees and scrub.

4 Fourth hind tarsal segment not longer than the first. 6.0–10.0 mm ***Galeruca* species**

Confirmatory characters: Pronotum and elytra black or buff. Elytra with more or less distinct longitudinal ridges.

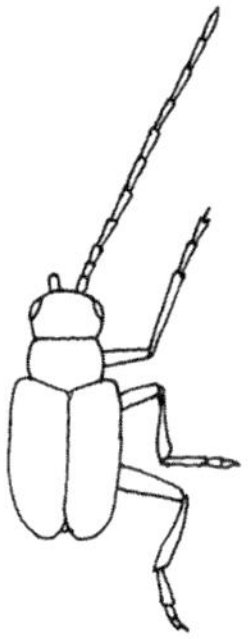

G.8 Galerucini long-legged species.

G.9 Galerucini shorter-legged species.

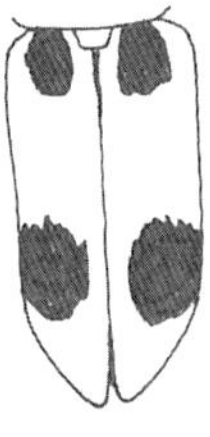

G.10 *Phyllobrotica* pattern.

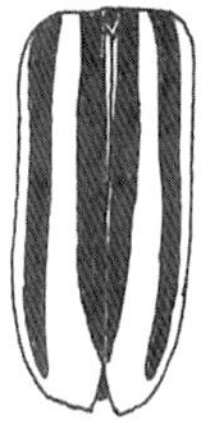

G.11 *Diabrotica* pattern.

– Fourth hind tarsal segment longer than the first. 3.0–7.0 mm **4**

5 Long antennae and slender legs (G.8). 3.8–5.0 mm **6**

– Without especially long antennae or slender legs (G.9). 3.0–7.0 mm **7**

6 Pronotum black or very dark, yellow or reddish-brown, never yellow with a black band at the rear. Elytra dark (usually black or blackish). Third antennal segment usually twice as long as second. 3.5–5.0 mm

***Luperus* species**

Confirmatory characters: Elytra hairless, though care is needed to see the hairs on *Calomicrus* below. Legs at least partly orange.

– Pronotum yellow with black band at the rear. Elytra yellow with black border along outer margin continuing along inner margin next to suture. Third antennal segment about as long as second. 3.0–4.5 mm

Calomicrus circumfusus

Confirmatory characters: Rear half of elytra sparsely hairy, though this can be very difficult to see. Legs black. Scarce and widely scattered in various habitats on Fabaceae, especially gorse.

7 Elytra yellowish with bold dark spots or lines (not only a dark suture). 5.0–7.0 mm **8**

[Only two species should key out here but both are distinctive; *Phyllobrotica quadrimaculata* and *Diabrotica virgifera*.]

– Yellowish, brownish or greyish. No dark spots or lines, but the suture may be dark. 3.0–7.0 mm **9**

8 Upper surface and appendages orange-yellow. Two black spots on each elytron (the front one may be missing) (G.10). Top of head black. 5.0–7.0 mm

Phyllobrotica quadrimaculata

Widespread, but not common, in various damp, mainly open, habitats on skullcaps (*Scutellaria*).

– Elytra shiny yellow with wide dark suture (may not reach the rear) and lateral longitudinal bands (G.11). Bands may be joined by a broad transverse band. Pronotum yellow, occasionally with small brown marks. Distinct shoulders. Head, legs and antennae dark (femora yellow beneath). 5.0–6.0 mm

Diabrotica virgifera (western corn rootworm)

Introduced pest of maize (*Zea mays*) found mainly near major airports.

9 Elytra hairless, smooth, irregularly punctured. Legs orange or yellowish, usually with at least femora darkened, sometimes black; if entirely orange, then elytra orange to red-brown with one or two black stripes or elongated spots. 3.7–6.0 mm

***Lochmaea* species**

Confirmatory characters: Elytra yellow-brown with lateral bump extending along the entire length, although this may be vague and hard to determine.

– Elytra finely or densely pubescent with hairs laid flat. Legs entirely orange or yellowish except in *Pyrrhalta* which often has a blackened dorsal ridge along the length of the tibia, or other dark marks. 3.0–7.0 mm 10

Confirmatory characters: Elytra with no lateral bump; at most a weak central bump.

10 Underside entirely reddish-brown. Head not especially small or short. 4.5–6.5 mm

Pyrrhalta viburni (viburnum leaf beetle)

Confirmatory characters: Pronotum and elytra yellow-brown, dull. Three dark longitudinal lines on pronotum. Pronotum only a little wider than head. Scutellum and elytral shoulders at least slightly darkened. Elytra dull with dense silky pubescence; punctures fine and dense. Head relatively large, narrowing slightly in front of and behind eyes; width across eyes only slightly less than pronotum. Widespread and fairly common on viburnums (*Viburnum*) in various habitats, including scattered records in Scotland which may indicate a northward range expansion although this is uncertain.

– Underside at least partly black. Head relatively small, narrowing slightly in front of and behind eyes; width across eyes considerably less than width of pronotum (G.12). 3.0–7.0 mm 11

G.12 Narrow eye width.

11 Antennal segment 3 longer than 4. 3.0–6.0 mm

***Galerucella* species**

Confirmatory characters: Brownish- or greyish-yellow. Elytra with fairly coarse punctures (usually with finer ones between); varies from entirely yellowish to mostly darkened, sometimes with dark longitudinal bands. Distinctly punctured pronotum much wider than head (shape of pronotum may vary). Front of head entirely yellow or orange.

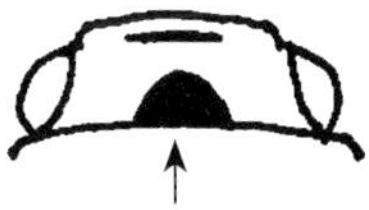

G.13 Head of female *Xanthogaleruca*.

G.14 Head of male *Xanthogaleruca*.

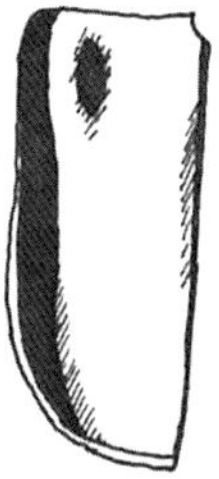

G.15 *Xanthogaleruca* elytra.

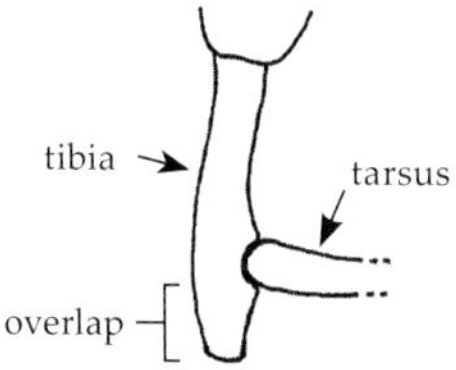

G.16 Leg of typical *Psylliodes*.

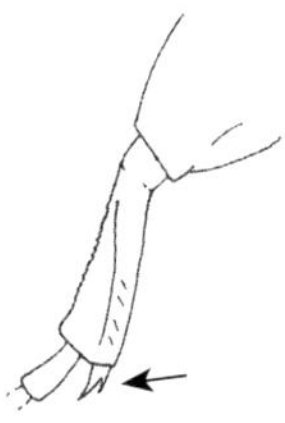

G.17 *Dibolia* tibial spurs.

– Antennal segments 3 and 4 approximately equal in length. 5.5–7.0 mm
Xanthogaleruca luteola (elm leaf beetle)

Confirmatory characters: Antennae blackish above. Pronotum yellow with three dark marks (can be variable) or a dark triangular spot. Head yellow with a dark spot in the female (G.13) and transverse band in the male (G.14). Elytra orange-yellow (sometimes yellow-brown or yellow-grey) with long dark longitudinal band from the shoulder (G.15). Occasional specimens, but not yet established. A major pest of elms.

Tribe Alticini

12 Hind tarsus joins tibia before the end, leaving a distinct overlap (G.16). Antennae with 10 segments. A large genus with many similar species. 2.0–4.5 mm
***Psylliodes* species**

– Hind tibia without an overlap at the tarsal joint. Antennae with 11 segments. 1.0–6.0 mm **13**

13 Tiny (1.0–1.5 mm), very convex (almost hemispherical) with head concealed from above. Black (usually with vague metallic sheen), antennae and legs reddish-brown. Last 3 antennal segments thickened
Mniophila muscorum

Scarce and widely scattered among mosses in woodland, parkland or moorland. Probably under-recorded due to its small size and tendency to be hidden among mosses, often in crevices in bark or similar.

– Not as above **14**

14 Spurs at end of hind tibiae wide and forked (G.17). 2.4–3.0 mm ***Dibolia cynoglossi***

Confirmatory characters: Metallic bronze or dark green, tibiae yellow-reddish, apical half of antennae dark brown. Hind femora greatly swollen – even more so than other flea beetles. Endangered (RDB1), on Lamiaceae in woodland rides, clearings and margins, on chalk hillsides and on coastal shingle. Recent records from only two sites in south-east England though may be under-recorded.

– Terminal spurs on hind tibiae with a single point. 1.0–6.0 mm **15**

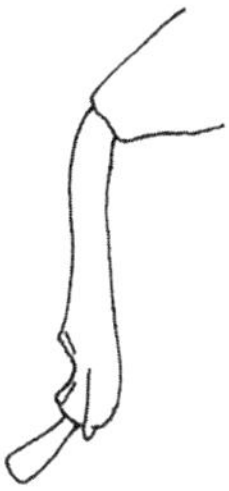

G.18 *Chaetocnema* tibia.

15 Distinct dent on middle and hind tibiae (G.18), with hairy upper margin (hairs may be difficult to see). 1.5–2.5 mm **_Chaetocnema_ species**

– Middle and hind tibiae without such a dent. 1.2–6.0 mm **16**

16 Pronotum without a groove parallel to the rear edge; at most, weak traces of short longitudinal furrow at the rear edge. If the pronotum has a pair of short furrows near the rear edge, then the dorsal body surface may vary in colour, but not entirely reddish-brownish. 1.0–6.0 mm **17**

Take care with *Hippuriphila modeeri* (couplet 21 below) as it can incorrectly key out here because the pronotal groove is very poorly defined. *H. modeeri* has a red-brown to bronze dorsal surface with paler elytral tips and is generally found on horsetails (*Equisetum*).

– Pronotum with a more or less distinct groove parallel to, and a little in front of, the rear edge, and/or a pair of short longitudinal furrows at the rear edge. If the groove is very shallow or faint, then either the elytra are densely pubescent, or the dorsal surface is entirely reddish-brownish and the groove ends with short furrows at right-angles to it. 1.2–5.5 mm **24**

G.19 Leg of typical *Longitarsus*.

17 First segment of hind tarsus at least half as long as hind tibia (G.19). In side view, apical spur of hind tibia as in G.17 above; when viewed from the rear, the spur is at the middle of lower edge. Wide range of colours and a large genus with many similar species. 1.0–4.0 mm **_Longitarsus_ species**

– First segment of hind tarsus less than half as long as hind tibia. Apical spur of hind tibia absent, small or variably placed. 1.4–5.0 mm **18**

18 Elytra randomly punctured or in irregular rows. 1.0–5.0 mm **19**

– Elytral punctures regularly striate (in rows), at least at the sides. 1.0–2.0 mm **22**

19 Round to oval, very convex (almost hemispherical). Upper surface brick-red to orange-red, sometimes yellowish or brownish. Antennae slender without thickened segments. 2.3–4.5 mm **_Sphaeroderma_ species**

– Oval, no more than moderately convex, may be slightly flattened **20**

20 Pronotum orange-red; elytra, head, legs and antennae black or dark brown. 5.0 mm

Luperomorpha xanthodera (rose flea beetle)

Introduced on garden plants (larvae are root-feeding) with occasional records from garden centres. Originally from China.

– Not this combination of colours **21**

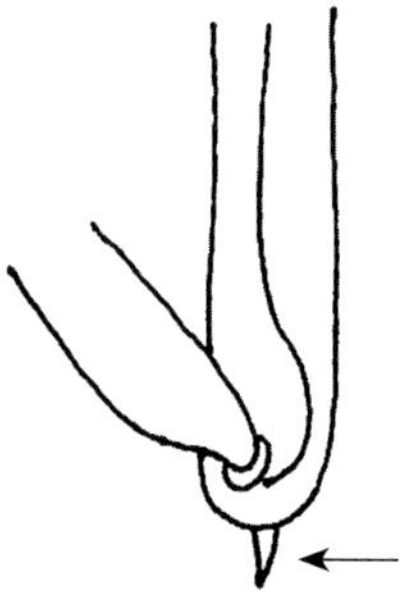

G.20 *Phyllotreta* tibial spur.

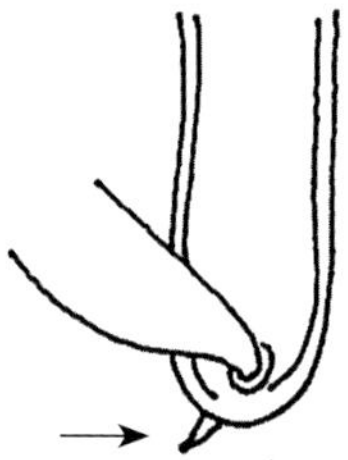

G.21 *Aphthona* tibial spur.

21 Bulges above antennal bases weakly convex, indistinct and not separated from front of the head by deep furrows. Front of the head above these weak bulges usually punctured. Hind tibiae without a longitudinal groove or 'gutter' on the dorsal side. Body dorsally dark or metallic, elytra sometimes with longitudinal yellow bands or other yellow markings. Apical spur of hind tibia at middle of lower edge (G.20). A large genus with many similar species. 1.4–3.5 mm

***Phyllotreta* species**

– Bulges above antennal bases usually well developed and separated from front of the head by deep furrows. Front of head above these bulges not punctured. Hind tibiae with a longitudinal groove or 'gutter' on the apical half of the dorsal side. Body dorsally yellow, brown, black, dark metallic blue or green. No yellow marks or bands on elytra. Apical spur of hind tibia at outer side of lower edge (G.21). 1.5–3.0 mm

***Aphthona* species**

Take care with the separation of *Phyllotreta* and *Aphthona*; some of the features may be difficult to see and differences can be subtle. However, careful observation and use of the range of features given should permit separation of these genera.

22 Short-oval, very convex, almost hemispherical. 2.2–3.0 mm ***Apteropeda* species**

– Body oval, not especially shortened, clearly not hemispherical **23**

23 Brownish, yellowish or reddish-yellow; not metallic. Head and pronotum rarely dark. Striae becoming faint towards tips of elytra. 1.8–2.0 mm

Lythraria salicariae

Confirmatory characters: Dark suture not reaching scutellum; where the darkening stops, diagonal lines sometimes run forward to shoulders forming faintly darkened triangles forward of this point. Scarce on loosestrifes (*Lysimachia*) and purple-loosestrife (*Lythrum salicaria*), sometimes chickweed-wintergreen (*Trientalis europaea*), in various habitats.

– Metallic dark bronze-green, dark brown or black. Striae formed of large punctures, not becoming faint towards tips. 1.0–2.0 mm **_Batophila_ species**

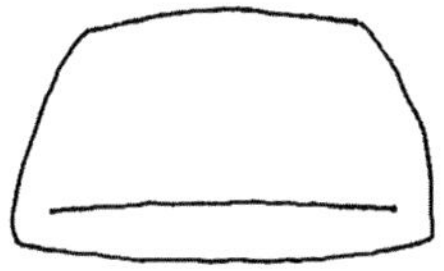

G.22 Pronotal groove.

24 Rear edge of pronotum with two short longitudinal furrows but no transverse groove (G.5 above) **25**

– Rear edge of pronotum with a transverse groove (e.g. G.3, G.4, G.22) **26**

25 Scutellar row of punctures reaches no more than one third the length of the suture. Head and pronotum red or reddish-brown. Elytra dark blue, green or violet, more-or-less metallic; punctures at least partly random. Base of pronotum curves inward to become narrower than front of elytra (G.23). 3.0–6.0 mm **_Podagrica_ species**

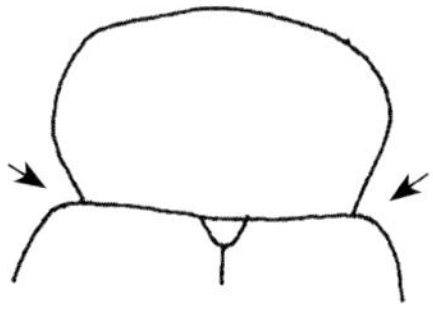

G.23 *Podagrica* pronotum.

– Scutellar row of punctures reaches more than half the length of the suture. Dorsally dark brown or blackish (tips of elytra sometimes black and brown) with metallic tinge. Elytral punctures striate. Rear edge of pronotum as wide as front of elytra. 1.8–2.8 mm **_Mantura_ species**

26 Pronotal groove extends almost to the sides of the pronotum; no short furrows at its ends (G.22). 2.8–5.5 mm **_Altica_ species**

Confirmatory characters: Metallic green, blue-green or blue. Body fairly flat. Elytra randomly punctured.

– Pronotal groove ends well before sides with short furrows joining base at right-angles (G.3, G.4). 1.2–5.5 mm **27**

G.24 *Hermaeophaga* coxal cavities.

27 Elytra randomly punctured. Front coxal cavities open at the rear (G.24). 2.3–3.0 mm **_Hermaeophaga mercurialis_**

Confirmatory characters: Blue-black; antennae and legs black except reddish-brown tarsi and antennal bases. Pronotum wide and convex; groove ends with short grooves joining base at right-angles, sometimes with a dent at each end (G.25). Body short-oval and very convex. Widespread and fairly common in southern

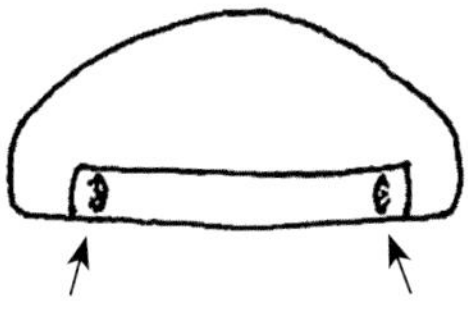

G.25 *Hermaeophaga* pronotal groove.

Britain, occasionally further north. On dog's mercury (*Mercurialis perennis*) in sunny woodland glades, hedgerows, chalk grassland, meadows, commons, heaths and quarries.

G.26 Coxal cavities closed at the rear.

– Elytral punctures striate, at least at the sides. Front coxal cavities closed at the rear (G.26). 1.2–5.5 mm **28**

28 Elytra hairy, elytral hairs erect and in rows. 1.2–2.0 mm ***Epitrix* species**

Confirmatory characters: Oval. Dark brown, black, sometimes yellow, rarely metallic. Pronotum hairy though this is often difficult to see.

– Elytra hairless, or if there is some pubescence it is only on the near-vertical sides. 1.8–5.5 mm **29**

29 Rear margin of bulges above antennal bases weakly separated from the remainder of the front of the head by a faint groove. 3.0–5.5 mm *Neocrepidodera* species

Confirmatory characters: Pale or reddish-brown, sometimes partly or wholly darker. Strong, unbroken transverse pronotal furrow (e.g. G.3). Top of head smooth or finely wrinkled, without coarse, dense punctures. Antennal segments 9 and 10 elongate (2–3 times as long as wide).

– Rear margin of bulges above antennal bases clearly separated from the remainder of the front of the head by a deep groove. 2.3–4.2 mm **30**

Take care with the separation of *Neocrepidodera* from the remaining genera; the depth of the groove on the head may be difficult to determine. However, careful observation and use of the range of features given should permit separation.

G.27 *Ochrosis* pronotal groove.

30 Dorsally unicolorous yellow or pale brown (no metallic reflection), head and prothorax rarely darkened; pronotum and elytra sometimes with a fine dark border. 2.1–2.5 mm ***Ochrosis ventralis***

Confirmatory characters: Transverse pronotal furrow weak with a central break (G.27). Elytral punctures becoming faint in the rear half. Bulges above antennal bases triangular with their rear edges separated from the remainder of the front of the head by a transverse furrow (G.28). Rare (RDB3), scattered and very local on various plants in lakesides, downland leys, coastal bays, cliffs and (probably) on disturbed ground, especially with free-draining soils.

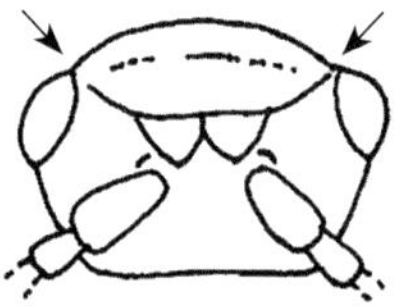

G.28 *Ochrosis* antennal bulges.

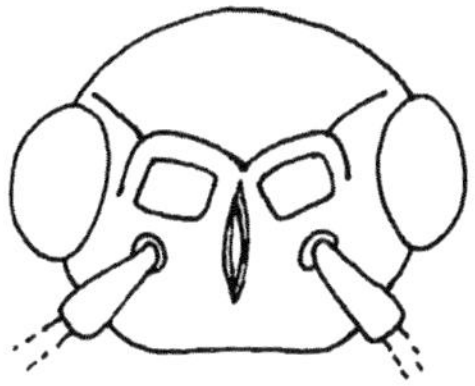

G.29 Antennal bulges.

– Dorsally bicolorous or with a metallic reflection **31**

Confirmatory characters: Elytral punctures not becoming faint in the rear half. Bulges above antennal bases roundish or transverse, surrounded by a deep furrow (G.29). Take care as the surrounding furrows join above the bulges, and may appear superficially similar to the transverse furrow in G.28 above. Also, the bulges may be more or less triangular in *Derocrepis rufipes* but still partly rounded and transverse on the inner and lower corners as in G.30 below.

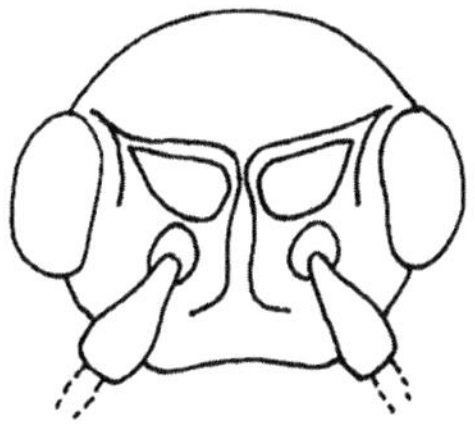

G.30 *Derocrepis* antennal bulges.

31 Head, pronotum, antennae and legs red or rusty. Elytra blackish to distinctly metallic blue, blue-green, bronze-green or blackish-green. 2.3–3.8 mm ***Derocrepis rufipes***

Confirmatory characters: Bulges above antennal bases large and rounded to triangular (G.30). Widespread and locally common on various Fabaceae in a range of habitats.

– Not this combination of colours **32**

32 Dorsal surface entirely metallic blue, green or bronze. Legs orange with hind femora darkened. 2.5–4.2 mm *Crepidodera* species

Bulges above antennal bases short and broad (e.g. G.29 above).

– Dorsally pitchy dark brown or black with weak bronze reflection; tips of elytra paler reddish-brownish. 1.8–2.5 mm ***Hippuriphila modeeri***

Widespread on horsetails (*Equisetum*) in various habitats.

Key H Subfamily Chrysomelinae

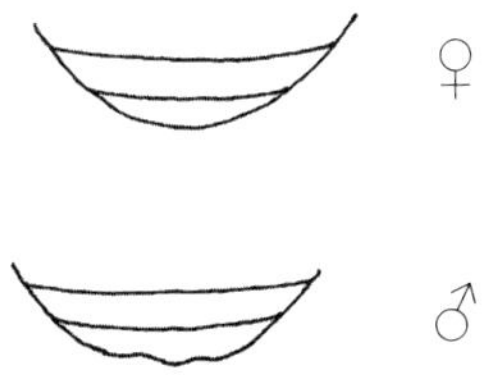

H.1 Tip of rear abdominal sternite.

Males usually have front tarsi with either one or the first three segments dilated, and this feature is not seen in females. Also, females have the tip of the rear abdominal sternite (underside plate) smoothly rounded; in males it is slightly indented or sinuous (H.1).

1 Looking from the side, inner edge of outward curve of the elytra with small bristles at or near the tip (H.2). Take care as this feature can be difficult to see. If missed, *Chrysolina* may key out as *Chrysomela, Gastrophysa, Plagiodera* or *Timarcha* hence confirmatory

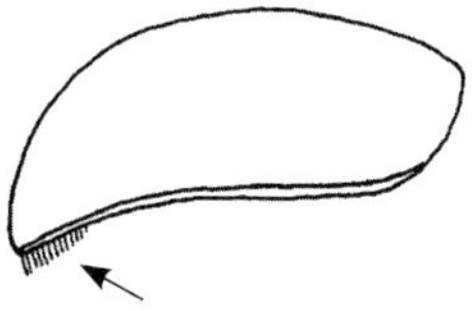

H.2 *Chrysolina* elytral bristles.

characters are particularly useful here. There is no need to separate the elytra to see the bristles. 4.5–11.0 mm *Chrysolina* species

Confirmatory characters: Tarsal claws without small tooth. Third hind tarsal segment only with a shallow dent at the tip. Last segment of maxillary palp at least as long as the preceding one. Measured along the midline, first abdominal sternite shorter than metasternum. Tibiae not especially dilated, and without a tooth at or near the apex. Pronotum only slightly narrower than elytra where they meet.

– Looking from the side, inner edge of elytra without small bristles at or near the tip. 2.5–18.0 mm **2**

2 Elytra with random punctures. 2.5–18.0 mm **3**

– Elytral punctures striate. 2.5–12.0 mm **6**

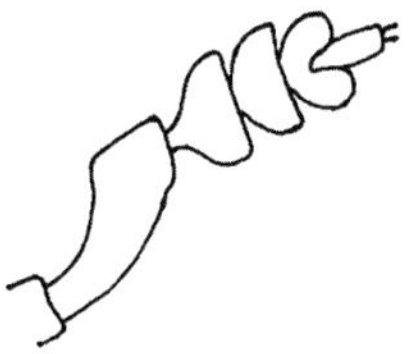

H.3 *Timarcha* tarsus.

3 Large (8.0–18.0 mm) beetles, darkly coloured (not metallic), wingless (elytra fused), nearly spherical, tarsal segments broadly spread, especially in males (H.3). Metasternum very short, constricted either side of the midline (H.4) ***Timarcha* species**

Take care with *Chrysolina violacea* which is superficially similar to *Timarcha goettingensis*, but has head, pronotum, elytra and underside metallic purple, and tarsi and palps orange brown and paler than the rest of the legs.

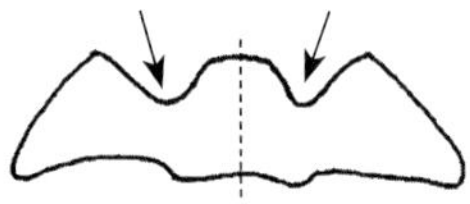

H.4 *Timarcha* metasternum.

– Winged (elytra not fused); may have some of the characters of *Timarcha* above, but not all. Metasternum longer without such strong constrictions (H.5). 2.5–12.0 mm **4**

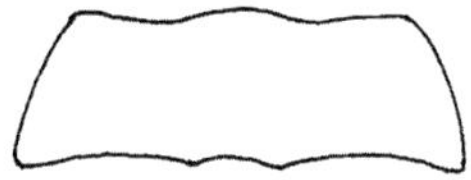

H.5 Metasternum (not *Timarcha*).

4 Elytra with a row of punctures along the rear third of the suture which is raised to form a narrow rim (H.6). 3.9–6.0 mm ***Gastrophysa* species**

Confirmatory characters: Slightly elongate, with elytra more-or-less parallel-sided. Pronotum somewhat bulging/domed. Female abdomen may become swollen as in the familiar 'green dock beetle' *G. viridula*, pushing aside the elytra to reveal black segments beneath. Brightly coloured and metallic. Front coxal cavities open and joined at the rear. Tooth at apex of middle and hind tibiae.

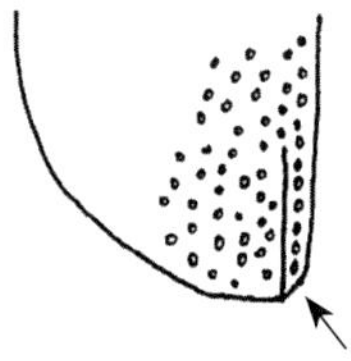

H.6 *Gastrophysa* elytral suture.

– Elytra without a row of punctures along the rear third

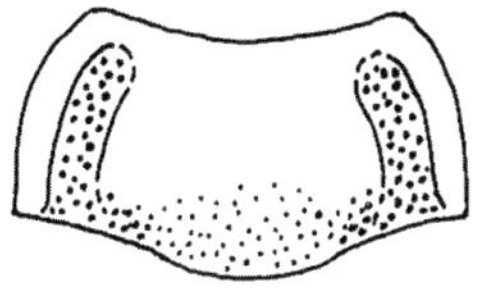

H.7 *Chrysomela* pronotum.

of the suture (at most a short row near the tip) which is not so distinctly raised as a rim. 2.5–12.0 mm **5**

5 Elongate to elongate-oval, moderately convex. Bumps to sides of pronotum separated from its upper surface by clear longitudinal dents (H.7) except in *Chrysomela aenea*. Pronotum considerably narrower than front of elytra. 6.3–12.0 mm ***Chrysomela* species**

– Round to rounded-oval, weakly convex. No bumps to sides of pronotum. Pronotum a little narrower than front of elytra. 2.5–4.8 mm ***Plagiodera versicolora***

Confirmatory characters: Dorsally metallic blue or green, rarely purplish or black. Head, pronotum and elytra usually concolourous (the same colour), occasionally elytra differ from head and pronotum. Underside, femora and tibiae black. Tip of elytra rounded with dimple (may be very shallow) in the apical angle (H.8). Elytra with narrow, more-or-less even rim, sides slightly indented above this rim. Elytra with distinct shoulders. Mostly in central and southern England, usually near water, on willows, especially crack-willow (*Salix fragilis*), sometimes far from water and/or on poplars and birches.

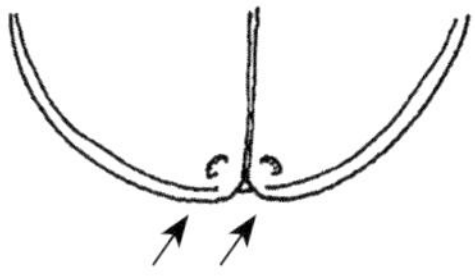

H.8 *Plagiodera* elytral tip.

H.9 Tarsal claws with appendages.

6 Claws without appendages at base. 3.0–6.0 mm **7**

– Claws with small appendages at base (H.9). 3.7–7.5 mm **9**

7 Metasternum with random punctures except for the front outer corner which is not punctured and is separated by a clear border (H.10). This corner area may be more or less densely microsculptured and occasionally a very few punctures can be found within (generally near the border), but clearly not the even puncturation covering the rest of the metasternum. 3.0–4.7 mm ***Phaedon* species**

Confirmatory characters: Rounded, domed oval, metallic. Side borders of elytra reach the tip. Elytra without yellow or reddish lateral stripe.

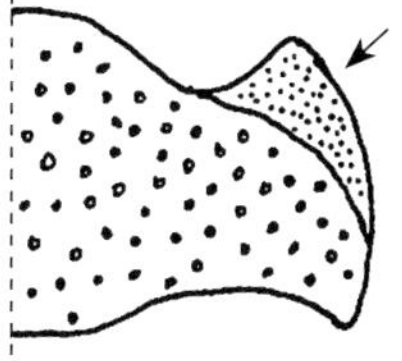

H.10 *Phaedon* metasternum.

– Metasternum with uniform coarse punctures, including front outer corner. 4.0–6.0 mm **8**

Confirmatory characters: Elytra sometimes with yellow or reddish lateral stripe.

8 Pronotum with rear margin (may be narrow). 4.0–6.0 mm ***Prasocuris* species**

Confirmatory characters: Elongate, parallel-sided with pronotum only a little narrower than elytra.

– Pronotum without a rear margin. 4.0–5.0 mm **_Hydrothassa_ species**

Confirmatory characters: Fairly elongate, metallic blue with yellowish edges to elytra and/or pronotum. Generally in or around wet habitats.

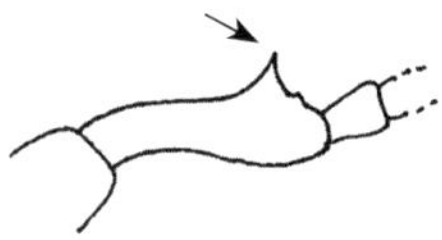

H.11 *Gonioctena* tibia.

9 Yellow, orange or reddish, with or without dark lines or spots. Middle and hind tibiae with a large thorn-like tooth on upper side near apex (H.11). There may be a similar, but often smaller, tooth on the front tibia. 3.7–7.5 mm **_Gonioctena_ species**

– Metallic but variable in colour. Legs and antennae brown or dark brown, with femora and apical antennal segments more likely to be dark. No tibial teeth. 3.7–5.0 mm **_Phratora_ species**

6.3 Systematic checklist of British and Irish species

This species list follows Duff (2012) with additional species (‡) discovered in the British Isles by the end of 2016. English common names are given for the minority of species that have them. However, these should be used with caution. For example, both *Batophila* species share the name 'raspberry flea beetle', while 'cabbage flea beetle' is used for *Phyllotreta cruciferae* but could equally well describe several similar species that feed on a range of brassicas. The name of each subfamily or species is followed by the name of the author or authors that first described them, together with the year in which the description was published. The scientific names are listed in standard systematic order with the species alphabetical within each genus. Note that the question mark in *Aphthona ?atratula* indicates uncertainty about the name and taxonomic status of this species.

Family: Megalopodidae

Subfamily	Scientific name	English name
Zeugophorinae Böving & Craighead, 1931	*Zeugophora flavicollis* (Marsham, 1802)	
	Zeugophora subspinosa (Fabricius, 1781)	
	Zeugophora turneri Power, 1863	

Family: Orsodacnidae

Subfamily	Scientific name	English name
Orsodacninae Thomson, 1859	*Orsodacne cerasi* (Linnaeus, 1758)	
	Orsodacne humeralis Latreille, 1804	

Family: Chrysomelidae

Subfamily	Scientific name	English name
Bruchinae Latreille, 1802	*Bruchus atomarius* (Linnaeus, 1761)	
	Bruchus brachialis Fåhraeus, 1839	
	Bruchus ervi Frölich, 1799	
	Bruchus loti Paykull, 1800	
	Bruchus pisorum (Linnaeus, 1758)	Pea beetle
	Bruchus rufimanus Boheman, 1833	Bean seed beetle
	Bruchus rufipes Herbst, 1783	
	Bruchidius cisti (Fabricius, 1775)	
	Bruchidius imbricornis (Panzer, 1795)‡	
	Bruchidius incarnatus (Boheman, 1833)	
	Bruchidius olivaceus (Germar, 1824)	
	Bruchidius siliquastri Delobel, 2007 ‡	
	Bruchidius varius (Olivier, 1795)	
	Bruchidius villosus (Fabricius, 1792)	
	Acanthoscelides obtectus (Say, 1831)	Dried bean beetle
	Callosobruchus chinensis (Linnaeus, 1758)	Adzuki beanseed beetle
	Callosobruchus maculatus (Fabricius, 1775)	Cowpea seed beetle
Donaciinae Kirby, 1837	*Macroplea appendiculata* (Panzer, 1794)	
	Macroplea mutica (Fabricius, 1792)	

Subfamily	Scientific name	English name
	Donacia aquatica (Linnaeus, 1758)	Zircon reed beetle
	Donacia bicolora Zschach, 1788	Two-tone reed beetle
	Donacia cinerea Herbst, 1784	
	Donacia clavipes Fabricius, 1792	
	Donacia crassipes Fabricius, 1775	Water-lily reed beetle
	Donacia dentata Hoppe, 1795	
	Donacia impressa Paykull, 1799	
	Donacia marginata Hoppe, 1795	
	Donacia obscura Gyllenhal, 1813	
	Donacia semicuprea Panzer, 1796	
	Donacia simplex Fabricius, 1775	
	Donacia sparganii Ahrens, 1810	
	Donacia thalassina Germar, 1811	
	Donacia versicolorea (Brahm, 1791)	
	Donacia vulgaris Zschach, 1788	
	Plateumaris bracata (Scopoli, 1772)	
	Plateumaris discolor (Panzer, 1795)	
	Plateumaris rustica (Kunze, 1818)	
	Plateumaris sericea (Linnaeus, 1758)	
Criocerinae Latreille, 1804	*Lema cyanella* (Linnaeus, 1758)	
	Oulema erichsoni (Suffrian, 1841)	
	Oulema melanopus (Linnaeus, 1758)	Cereal leaf beetle
	Oulema obscura (Stephens, 1831)	
	Oulema rufocyanea (Suffrian, 1847) 677	
	Oulema septentrionis (Weise, 1880)	
	Crioceris asparagi (Linnaeus, 1758)	Asparagus beetle
	Crioceris macilenta Weise, 1880 ‡	
	Lilioceris lilii (Scopoli, 1763)	Lily beetle
Cryptocephalinae Gyllenhal, 1813	*Labidostomis tridentata* (Linnaeus, 1758)	
	Clytra laeviuscula Ratzeburg, 1837	
	Clytra quadripunctata (Linnaeus, 1758)	Four-spotted leaf beetle
	Smaragdina affinis (Illiger, 1794)	
	Smaragdina salicina (Scopoli, 1763) ‡	
	Cryptocephalus aureolus Suffrian, 1847	
	Cryptocephalus biguttatus (Scopoli, 1763)	

Subfamily	Scientific name	English name
	Cryptocephalus bilineatus (Linnaeus, 1767)	
	Cryptocephalus bipunctatus (Linnaeus, 1758)	
	Cryptocephalus coryli (Linnaeus, 1758)	Hazel pot beetle
	Cryptocephalus decemmaculatus (Linnaeus, 1758)	Ten-spotted pot beetle
	Cryptocephalus exiguus Schneider, D.H., 1792	Pashford pot beetle
	Cryptocephalus frontalis Marsham, 1802	
	Cryptocephalus fulvus (Goeze, 1777)	
	Cryptocephalus hypochaeridis (Linnaeus, 1758)	
	Cryptocephalus labiatus (Linnaeus, 1761)	
	Cryptocephalus moraei (Linnaeus, 1758)	
	Cryptocephalus nitidulus Fabricius, 1787	Shining pot beetle
	Cryptocephalus parvulus Müller, O.F., 1776	
	Cryptocephalus primarius Harold, 1872	Rock-rose pot beetle
	Cryptocephalus punctiger Paykull, 1799	Blue pepper-pot beetle
	Cryptocephalus pusillus Fabricius, 1777	
	Cryptocephalus querceti Suffrian, 1848	Oak pot beetle
	Cryptocephalus sexpunctatus (Linnaeus, 1758)	Six-spotted pot beetle
	Cryptocephalus violaceus Laicharting, 1781	
Lamprosomatinae Lacordaire, 1848	*Oomorphus concolor* (Sturm, 1807)	
Eumolpinae Hope, 1840	*Bromius obscurus* (Linnaeus, 1758)	
Chrysomelinae Latreille, 1802	*Timarcha goettingensis* (Linnaeus, 1758)	Lesser bloody-nosed beetle
	Timarcha tenebricosa (Fabricius, 1775)	Bloody-nosed beetle
	Chrysolina americana (Linnaeus, 1758)	Rosemary beetle
	Chrysolina banksii (Fabricius, 1775) 682	
	Chrysolina brunsvicensis (Gravenhorst, 1807)	
	Chrysolina cerealis (Linnaeus, 1767)	Rainbow leaf beetle
	Chrysolina coerulans (Scriba, 1791)	
	Chrysolina fastuosa (Scopoli, 1763)	Dead-nettle leaf beetle

Subfamily	Scientific name	English name
	Chrysolina graminis (Linnaeus, 1758)	Tansy beetle
	Chrysolina haemoptera (Linnaeus, 1758)	Plantain leaf beetle
	Chrysolina herbacea (Duftschmid, 1825)	Mint leaf beetle
	Chrysolina hyperici (Forster, 1771)	
	Chrysolina latecincta (Demaison, 1896) subspecies *intermedia* (Franz, 1938)	
	Chrysolina marginata (Linnaeus, 1758)	
	Chrysolina oricalcia (Müller, O.F., 1776)	
	Chrysolina polita (Linnaeus, 1758)	Knotgrass leaf beetle
	Chrysolina sanguinolenta (Linnaeus, 1758)	Toadflax leaf beetle
	Chrysolina staphylaea (Linnaeus, 1758)	
	Chrysolina sturmi (Westhoff, 1882)	
	Chrysolina varians (Schaller, 1783)	
	Gastrophysa polygoni (Linnaeus, 1758)	
	Gastrophysa viridula (De Geer, 1775)	Green dock beetle
	Phaedon armoraciae (Linnaeus, 1758)	
	Phaedon cochleariae (Fabricius, 1792)	Water-cress beetle
	Phaedon concinnus Stephens, 1831	
	Phaedon tumidulus (Germar, 1824)	Celery leaf beetle
	Hydrothassa glabra (Herbst, 1783)	
	Hydrothassa hannoveriana (Fabricius, 1775)	
	Hydrothassa marginella (Linnaeus, 1758)	
	Prasocuris junci (Brahm, 1790)	Brooklime leaf beetle
	Prasocuris phellandrii (Linnaeus, 1758)	
	Plagiodera versicolora (Laicharting, 1781)	
	Chrysomela aenea Linnaeus, 1758	
	Chrysomela populi Linnaeus, 1758	Red poplar leaf beetle
	Chrysomela saliceti Suffrian, 1851 ‡	
	Chrysomela tremula Fabricius, 1787	
	Chrysomela vigintipunctata (Scopoli, 1763) ‡	
	Gonioctena decemnotata (Marsham, 1802)	
	Gonioctena olivacea (Forster, 1771)	Broom leaf beetle
	Gonioctena pallida (Linnaeus, 1758)	

Subfamily	Scientific name	English name
	Gonioctena viminalis (Linnaeus, 1758)	
	Phratora laticollis Suffrian, 1851	
	Phratora polaris Schneider, J.S., 1886	
	Phratora vitellinae (Linnaeus, 1758)	Brassy willow beetle
	Phratora vulgatissima (Linnaeus, 1758)	Blue willow beetle
	Paropsisterna selmani Reid & de Little, 2013 ‡	Eucalyptus leaf beetle
Galerucinae Latreille, 1802	*Galerucella calmariensis* (Linnaeus, 1767)	
	Galerucella lineola (Fabricius, 1781)	Brown willow beetle
	Galerucella nymphaeae (Linnaeus, 1758)	
	Galerucella pusilla (Duftschmid, 1825)	
	Galerucella sagittariae (Gyllenhal, 1813)	
	Galerucella tenella (Linnaeus, 1761)	
	Pyrrhalta viburni (Paykull, 1799)	Viburnum leaf beetle
	Xanthogaleruca luteola (Müller, O.F., 1766)	Elm leaf beetle
	Galeruca laticollis (Sahlberg, C.R., 1838)	
	Galeruca tanaceti (Linnaeus, 1758)	
	Lochmaea caprea (Linnaeus, 1758)	Willow leaf beetle
	Lochmaea crataegi (Forster, 1771)	Hawthorn leaf beetle
	Lochmaea suturalis (Thomson, C.G., 1866)	Heather beetle
	Diabrotica virgifera LeConte, 1858	Western corn rootworm
	Phyllobrotica quadrimaculata (Linnaeus, 1758)	Skullcap leaf beetle
	Luperus flavipes (Linnaeus, 1767)	
	Luperus longicornis (Fabricius, 1781)	
	Calomicrus circumfusus (Marsham, 1802)	
	Agelastica alni (Linnaeus, 1758)	Alder leaf beetle
	Sermylassa halensis (Linnaeus, 1767)	
	Luperomorpha xanthodera (Fairmaire, 1888)	
	Phyllotreta atra (Fabricius, 1775)	
	Phyllotreta consobrina (Curtis, 1837)	
	Phyllotreta cruciferae (Goeze, 1777)	Cabbage flea beetle
	Phyllotreta diademata Foudras, 1860	

Subfamily	Scientific name	English name
	Phyllotreta exclamationis (Thunberg, 1784)	
	Phyllotreta flexuosa (Illiger, 1794)	
	Phyllotreta nemorum (Linnaeus, 1758)	Large striped flea beetle
	Phyllotreta nigripes (Fabricius, 1775)	Turnip flea beetle
	Phyllotreta nodicornis (Marsham, 1802)	
	Phyllotreta ochripes (Curtis, 1837)	
	Phyllotreta punctulata (Marsham, 1802)	
	Phyllotreta striolata (Fabricius, 1803)	Striped flea beetle
	Phyllotreta tetrastigma (Comolli, 1837)	
	Phyllotreta undulata Kutschera, 1860	Small striped flea beetle
	Phyllotreta vittula (Redtenbacher, 1849)	Barley flea beetle
	Aphthona ?atratula Allard, 1859	
	Aphthona atrocaerulea (Stephens, 1829)	
	Aphthona euphorbiae (Schrank, 1781)	Large flax flea beetle
	Aphthona herbigrada (Curtis, 1837)	
	Aphthona lutescens (Gyllenhal, 1808)	
	Aphthona melancholica Weise, 1888	
	Aphthona nigriceps (Redtenbacher, 1842)	
	Aphthona nonstriata (Goeze, 1777)	Iris flea beetle
	Aphthona pallida (Bach, 1856)	
	Longitarsus absynthii Kutschera, 1862	
	Longitarsus aeneicollis (Faldermann, 1837)	
	Longitarsus aeruginosus (Foudras, 1860)	
	Longitarsus agilis (Rye, 1868)	
	Longitarsus anchusae (Paykull, 1799)	
	Longitarsus atricillus (Linnaeus, 1761)	
	Longitarsus ballotae (Marsham, 1802)	
	Longitarsus brunneus (Duftschmid, 1825)	
	Longitarsus curtus (Allard, 1860)	
	Longitarsus dorsalis (Fabricius, 1781)	
	Longitarsus exoletus (Linnaeus, 1758)	
	Longitarsus ferrugineus (Foudras, 1860)	Mint flea beetle
	Longitarsus flavicornis (Stephens, 1831)	

Subfamily	Scientific name	English name
	Longitarsus ganglbaueri Heikertinger, 1912	
	Longitarsus gracilis Kutschera, 1864	
	Longitarsus holsaticus (Linnaeus, 1758)	
	Longitarsus jacobaeae (Waterhouse, G.R., 1858)	
	Longitarsus kutscherae (Rye, 1872)	
	Longitarsus longiseta Weise, 1889	
	Longitarsus luridus (Scopoli, 1763)	
	Longitarsus lycopi (Foudras, 1860)	
	Longitarsus melanocephalus (De Geer, 1775)	
	Longitarsus membranaceus (Foudras, 1860)	
	Longitarsus minusculus (Foudras, 1860) ‡	
	Longitarsus nasturtii (Fabricius, 1792)	
	Longitarsus nigerrimus (Gyllenhal, 1827)	
	Longitarsus nigrofasciatus (Goeze, 1777)	
	Longitarsus obliteratoides Gruev, 1973	
	Longitarsus obliteratus (Rosenhauer, 1847)	
	Longitarsus ochroleucus (Marsham, 1802)	
	Longitarsus parvulus (Paykull, 1799)	Flax flea beetle
	Longitarsus pellucidus (Foudras, 1860)	
	Longitarsus plantagomaritimus Dollman, 1912	
	Longitarsus pratensis (Panzer, 1794)	
	Longitarsus quadriguttatus (Pontoppidan, 1763)	
	Longitarsus reichei (Allard, 1860)	
	Longitarsus rubiginosus (Foudras, 1860)	
	Longitarsus rutilus (Illiger, 1807)	
	Longitarsus strigicollis Wollaston, 1864 *	
	Longitarsus succineus (Foudras, 1860)	Chrysanthemum flea beetle
	Longitarsus suturellus (Duftschmid, 1825)	

Subfamily	Scientific name	English name
	Longitarsus symphyti Heikertinger, 1912	
	Longitarsus tabidus (Fabricius, 1775)	
	Altica brevicollis Foudras, 1860	
	Altica carinthiaca Weise, 1888	
	Altica helianthemi (Allard, 1859)	
	Altica longicollis (Allard, 1860)	
	Altica lythri Aubé, 1843	
	Altica oleracea (Linnaeus, 1758)	
	Altica palustris Weise, 1888	
	Hermaeophaga mercurialis (Fabricius, 1792)	Dog's-mercury flea beetle
	Batophila aerata (Marsham, 1802)	Raspberry flea beetle
	Batophila rubi (Paykull, 1799)	Raspberry flea beetle
	Lythraria salicariae (Paykull, 1800)	Loosestrife flea beetle
	Ochrosis ventralis (Illiger, 1807)	
	Neocrepidodera ferruginea (Scopoli, 1763)	Wheat flea beetle
	Neocrepidodera impressa (Fabricius, 1801)	
	Neocrepidodera transversa (Marsham, 1802)	
	Derocrepis rufipes (Linnaeus, 1758)	
	Hippuriphila modeeri (Linnaeus, 1761)	Horsetail flea beetle
	Crepidodera aurata (Marsham, 1802)	Willow flea beetle
	Crepidodera aurea (Fourcroy, 1785)	
	Crepidodera fulvicornis (Fabricius, 1792)	
	Crepidodera nitidula (Linnaeus, 1758)	
	Crepidodera plutus (Latreille, 1804)	
	Epitrix atropae Foudras, 1860	Belladonna flea beetle
	Epitrix pubescens (Koch, J.D.W., 1803)	
	Podagrica fuscicornis (Linnaeus, 1767)	
	Podagrica fuscipes (Fabricius, 1775)	Mallow flea beetle
	Mantura chrysanthemi (Koch, J.D.W., 1803)	
	Mantura matthewsii (Curtis, 1833)	
	Mantura obtusata (Gyllenhal, 1813)	
	Mantura rustica (Linnaeus, 1767)	
	Chaetocnema aerosa (Letzner, 1847)	

Subfamily	Scientific name	English name
	Chaetocnema arida Foudras, 1860	
	Chaetocnema concinna (Marsham, 1802)	Mangold flea beetle
	Chaetocnema confusa (Boheman, 1851)	
	Chaetocnema hortensis (Fourcroy, 1785)	
	Chaetocnema picipes Stephens, 1831	
	Chaetocnema sahlbergii (Gyllenhal, 1827)	
	Chaetocnema subcoerulea (Kutschera, 1864)	
	Sphaeroderma rubidum (Graëlls, 1858)	
	Sphaeroderma testaceum (Fabricius, 1775)	
	Apteropeda globosa (Illiger, 1794)	
	Apteropeda orbiculata (Marsham, 1802)	
	Apteropeda splendida Allard, 1860	
	Mniophila muscorum (Koch, J.D.W., 1803)	Moss flea beetle
	Dibolia cynoglossi (Koch, J.D.W., 1803)	
	Psylliodes affinis (Paykull, 1799)	Potato flea beetle
	Psylliodes attenuata (Koch, J.D.W., 1803)	Hop flea beetle
	Psylliodes chalcomera (Illiger, 1807)	
	Psylliodes chrysocephala (Linnaeus, 1758)	Cabbage-stem flea beetle
	Psylliodes cucullata (Illiger, 1807) 707	
	Psylliodes cuprea (Koch, J.D.W., 1803)	
	Psylliodes dulcamarae (Koch, J.D.W., 1803)	
	Psylliodes hyoscyami (Linnaeus, 1758)	Henbane flea beetle
	Psylliodes laticollis Kutschera, 1860	
	Psylliodes luridipennis Kutschera, 1864	Lundy cabbage flea beetle
	Psylliodes luteola (Müller, O.F., 1776)	
	Psylliodes marcida (Illiger, 1807)	
	Psylliodes napi (Fabricius, 1792)	
	Psylliodes picina (Marsham, 1802)	
	Psylliodes sophiae Heikertinger, 1914	Flixweed flea beetle
Cassidinae Gyllenhal, 1813	*Pilemostoma fastuosa* (Schaller, 1783)	
	Hypocassida subferruginea (Schrank, 1776)	

Subfamily	Scientific name	English name
	Cassida denticollis Suffrian, 1844	
	Cassida flaveola Thunberg, 1794	Pale tortoise beetle
	Cassida hemisphaerica Herbst, 1799	
	Cassida murraea Linnaeus, 1767	Fleabane tortoise beetle
	Cassida nebulosa Linnaeus, 1758	
	Cassida nobilis Linnaeus, 1758	
	Cassida prasina Illiger, 1798	
	Cassida rubiginosa Müller, O.F., 1776	Thistle tortoise beetle
	Cassida sanguinosa Suffrian, 1844	
	Cassida vibex Linnaeus, 1767	
	Cassida viridis Linnaeus, 1758	Green tortoise beetle
	Cassida vittata de Villers, 1789	Banded tortoise beetle

* *Longitarsus fowleri* Allen, 1967 has recently been synonymised with (and therefore should be known as) *Longitarsus strigicollis* (Cox, 2015).

7 Study techniques and materials

Leaf beetles can be found associated with vegetation in most terrestrial habitats in Britain and Ireland, as well as freshwater habitats with marginal and emergent plants. None live in strictly saline habitats, but saltmarshes, coastal dunes, estuaries, coastal shingle and coastal cliffs all support a range of species. Bruchids may be found indoors, especially in premises where seeds of their foodplants are stored.

coleopterist
someone who studies beetles (Coleoptera)

Leaf beetles may therefore be found anywhere with vegetation. Although 'high quality' sites such as nature reserves are popular locations for coleopterists to visit, other sites should not be overlooked. For example, waste ground, urban and brownfield sites may not be as attractive as many SSSIs, but can support a diverse fauna, including scarce and rare species (Gibson, 1998) and many such sites are under-recorded. Conversely, while ancient trees in woodland, parkland and wood-pasture are of great importance for many insect species, leaf beetles are not generally associated with ancient trees or dead wood, although the rare *Cryptocephalus querceti* and *Smaragdina affinis* are known from such situations. So, while woodlands, wetlands, chalk grassland, saltmarsh and the margins of water bodies are likely to support a rich leaf beetle fauna, any location may be worth investigating as most species are highly mobile and may colonise locations where their foodplants are present.

Studying insects often means taking specimens as an unavoidable part of the identification process, and it is important to ensure this is done with conservation in mind. In some cases, identification may be possible in the field, in which case the specimens can be released, but usually they need to be killed and a microscope used. To some extent, this depends on the location; for example, *Timarcha* found in Britain can be readily identified as there are only two species and these are not difficult to separate with practice. However, in continental Europe there are several more species as well as a number of recognised subspecies and specimens may be required. Similarly, if collecting in locations not covered by comprehensive and up-to-date identification literature, it is likely that specimens will be needed for comparison with collections. The rapid development of digital photography means that images may occasionally be sufficiently detailed to permit identification, but this is generally not the case.

In general, collecting specimens does not cause significant harm to the population or habitat being sampled, but care needs to be taken, especially when studying juvenile stages or rare species. The 'Code of Conduct' for insect collecting (Invertebrate Link, 2002) should be followed and for some locations, site owners or managers may need to be consulted and permission granted.

The following sections cover curatorial aspects of leaf beetle study. This is essentially a brief summary of a complex area where skills are developed through practical experience, but more detail is given in the book by Cooter & Barclay (2006), which is highly recommended.

7.1 Collecting, preserving and storing

Collecting specimens by hand-searching is possible with minimal field equipment (see Chapter 8 for details of equipment suppliers) i.e. a pooter, small clear glass or plastic specimen containers and a hand lens. In some cases, beetles can be picked up by hand, with a small paintbrush, or knocked into a container, but often a pooter will be needed because flea beetles' leaping ability helps them evade entomologists as well as predators! A number of locations are particularly suited to this approach, such as flood debris, under stones, and the area beneath the rosettes and around the roots of plants, especially larger herbaceous species. The final section of this chapter provides an annotated list of plant groups to help with searching. In aquatic habitats, a plant with beetles on it can be quickly submerged and the beetles which then float to the surface collected by hand.

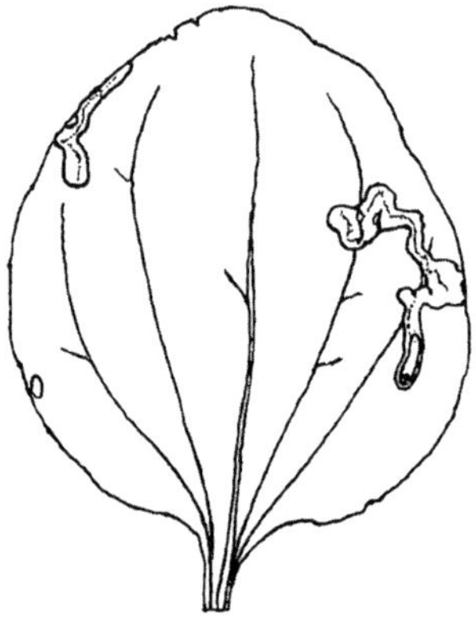

Fig. 7.1 *Apteropeda* leaf mine.

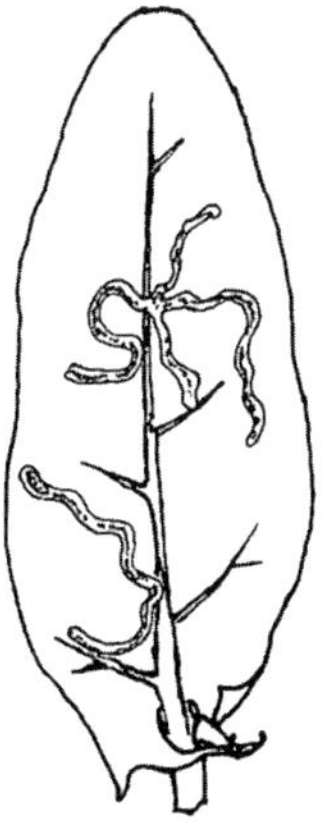

Fig. 7.2 *Mantura* leaf mine.

Fig. 7.3 *Phyllotreta* leaf mine.

Looking for leaf mines can help locate larvae of a number of species, especially in the Alticini, but also Megalopodidae. For example:

- *Apteropeda orbiculata* on a wide range of plants including plantains and dead-nettles (Fig. 7.1).
- *Mantura* species on various plants, especially docks and sorrels, with *Mantura rustica* also using knotgrass, and *Mantura matthewsi* being found on rock-roses (Fig. 7.2).
- *Phyllotreta nemorum* is usually associated with a wide range of Brassicaceae (Fig. 7.3).
- *Sphaeroderma* species are found on a range of Asteraceae, while *Dibolia cynoglossi* uses many species

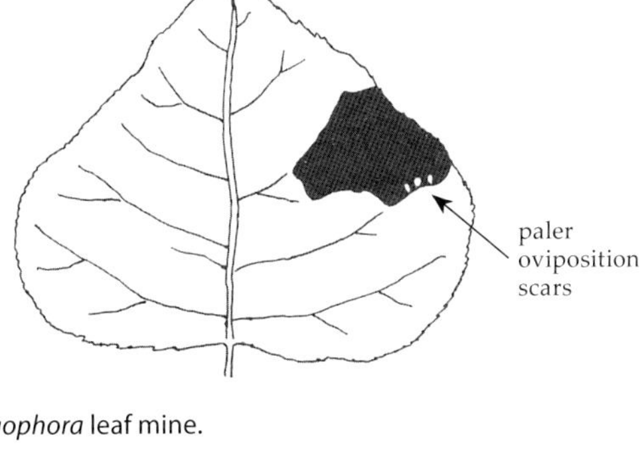

Fig. 7.4 *Zeugophora* leaf mine.

of dead-nettles. In these species, the mines eventually fill the whole leaf.

- *Zeugophora* species on poplars and willows. (Fig. 7.4).

In some other species, feeding damage on leaves is distinctive enough to indicate the presence of the beetle. A good example of this is *Hermaeophaga mercurialis* which feeds only on dog's-mercury *Mercurialis perennis* and creates 'shot-holes' in the leaves. Such feeding marks (Fig. 7.5) are caused by many Alticini and can be useful indicators of species presence when the beetle has a restricted range of foodplants.

Fig. 7.5 Shotholes due to alticine feeding activity.

Simple tools such as a household table fork may help prise apart roots to find beetles hiding among new growth. Other locations require more destructive searching such as beneath bark and inside bulrush leaves where beetles may hibernate. Using a series of sieves with different mesh sizes, starting with the largest, soil and leaf litter can be sorted to find larvae and pupae as well as overwintering adults. This may find specimens that would otherwise be missed but is time consuming as a large amount of material is likely to be needed to find a small number of beetles.

Similarly, grass tussocks can be cut away 2 or 3 cm below ground level, shaken out over a pale sheet (the pooter will be needed) and dismantled to find beetles. As with any destructive method, the importance of obtaining records needs to be balanced against maintenance of the habitat, though of course the information gained can be used to inform conservation efforts and increase knowledge of the species (Sections 1.4 and 5.1). However, hand-searching is limited and will miss beetles in locations such as trees and bushes, or among long grass. Also, many leaf beetles are small and few are brightly coloured, and so are easy to miss. Care is needed not to damage beetles, especially if they are small, and if eggs or pupae are taken, the section of vegetation they are attached to should be taken with them still attached.

Sweeping is a technique where one sweeps a net from side to side while walking slowly through vegetation. This allows a larger area to be sampled than would be possible by trying to hand-search. Sweep nets are sturdier than butterfly nets, which have a fine mesh that is easily torn on tough vegetation. This does not work so well for aquatic species and a sturdier net is needed, such as a kick-sampling net with a strong handle and frame. This must be able to tolerate the weight of water and plant material that is collected and is useful when sampling donaciines. Another version is a long-handled clap net which can be closed from a distance and is used to reach and sample inaccessible locations.

Beating involves using a stick to sharply tap tree and shrub branches to dislodge beetles which fall onto a beating tray – a piece of white cloth held horizontally on a frame. They can then be collected from the tray in the same way as hand-searching. Alternatives to a beating tray include an old umbrella (preferably a plain, pale-coloured one), or a piece of white or pale cloth spread on the ground.

Other methods involve trapping and the simplest of these are pitfall and water traps. Pitfall traps consist of containers such as glass jars or plastic pots around 10–15 cm high which are dug into the soil so the top is at ground level (Fig. 7.6). These catch insects that walk into them and if checked approximately daily can either be dry, or contain a small amount of water with a few drops of detergent to

Fig. 7.6 Pitfall trap.

break the surface tension. Other substances should be avoided due to their high evaporation rate (e.g. alcohols) or toxicity (e.g. formalin). If high levels of rainfall are likely, the trap should have a cover raised a few centimetres above it, or if it is a dry plastic trap, small drainage holes can be made. An amended version known as a water trap can be used to catch species that are attracted to yellow, especially in dry locations, for example brassica-feeding *Phyllotreta* species. A shallow yellow dish or pan is placed on the ground or sunk like a pitfall trap and contains water with detergent as above. Again, this should be checked and emptied daily if possible. These fluid traps will catch a variety of insects, but are small and unlikely to have a significant impact on populations. As with any sampling method, if non-chrysomelids are caught, it is worth identifying them and sending the results to the relevant recording scheme or Biological Records Centre if possible.

A variety of more specialised or sophisticated trapping equipment can be bought from entomological and ecological suppliers, or in some cases can be home-made. They tend to be used for large-scale sampling, surveying or research, and care is needed as some are indiscriminate and can catch and kill large numbers of insects. The detailed use of these is beyond the scope of this work, but the main types are as follows:

- Suction sampler, such as the D-VAC, Vortis sampler, or a modified motorised garden leaf collector.
- Suction trap, which uses a fan to draw air through a net suspended above the ground to catch flying insects.
- Fogging uses a fine mist of insecticide to sample tree canopies with a remote-controlled sprayer suspended on a rope and collecting sheets placed beneath to catch insects.
- Malaise traps are fine tent-like structures used to catch flying insects by funnelling them into a collecting jar with a killing agent (Fig. 7.7).
- Flight interception traps are vertical sheets of clear glass or plastic suspended over a tray of water which capture flying insects that hit the sheet. Alternatively, fine black mesh can be used with a tray or trough placed to catch the insects. Smaller versions may be hung in trees with some consisting of two sheets intersecting at 90° to form four 'vanes'.

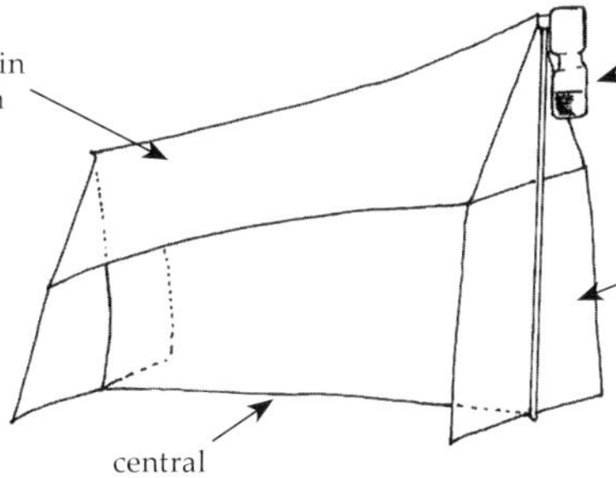

Fig. 7.7 Malaise trap.

- Light traps such as those used by to catch nocturnal moths.
- Emergence cages are mesh bags tied over twigs or small plants to capture adults when they emerge.

Whatever collection method is used, the resulting catch needs to be processed in order to be identified. In some cases, beetles can be identified alive and then returned to their original habitat. Sometimes, being refrigerated for several minutes may temporarily slow their movements enough to allow identification using external features; if so, they can again be released. However, it is likely that specimens will need to be killed to be studied.

This is usually done by putting a few drops of ethyl acetate onto an absorbent material such as thick tissue or filter paper and placing this into a sealed glass jar along with the beetles. Ideally the beetles need to be kept out of direct contact with the fluid, for example by placing them in a smaller tube plugged loosely to prevent them escaping while allowing vapour to enter. Avoid inhaling the vapour yourself. Alternatively, beetles may be killed by placing them in a freezer for around 30 minutes.

Once dead, beetles should be mounted as soon as possible as they become brittle when dry. If they have become stiff and brittle, they can be relaxed by placing them in a box lined with tissue or wadding soaked in water. The box needs a tight-fitting lid and is left for a few hours in a warm place. This softens the muscles and connective tissue so that the appendages can be moved for setting or the abdomen dissected. Commercial relaxing fluids are available, but as water works well, they should not be needed.

Beetles can then be mounted by either carding or pinning

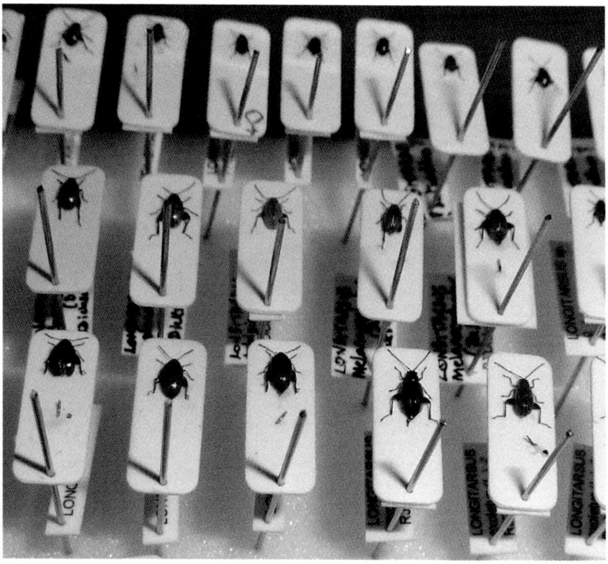

Fig. 7.8 Pinned specimens.

(Fig. 7.8). Die-stamped cards, including card points for small species, are available cheaply enough not to need to cut your own, but if you wish to, 3- or 4-ply Bristol board should be used and is available from art suppliers. Whichever glue you choose, it needs to be water-soluble, clear, durable, matt when dry and non-staining or darkening over time. There are many options, but popular ones include clear starch-based glue or similar office and stationery adhesive, and Watkins & Doncaster's mounting glue. A small amount of dilute glue is brushed onto the card and the underside of the beetle placed on it. Excess glue, or glue that is too viscous, should be avoided as the surface microsculpture, pubescence and other fine details need to remain visible, and the appendages, including legs, antennae and palpi neatly arranged. Fine forceps and paintbrushes may be used to manipulate fine structures, especially of small beetles. The card can then be pinned onto white Plastazote foam (see Chapter 8 for details of equipment suppliers), and ideally a pinning stage used to ensure even-spacing of cards and data labels.

pinning stage
a metal or wooden block used to position specimens and labels on pins. Available from equipment suppliers listed in Chapter 8

Although British coleopterists have traditionally carded their specimens, my preference is for pinning unless specimens are especially tiny. For larger beetles, an entomologists' pin is pushed vertically through the right elytron just behind the front edge and about a third of the elytron width from the suture where the elytra meet. As the pin is slowly and gently pushed so that it emerges through the ventral surface, check the point where it appears to ensure it does not pierce or dislodge the coxa. If it does, withdraw the point and push through again avoiding the coxa. If the specimen is small, a short, fine micropin is used instead. This attaches the beetle near one end of a narrow strip of Plastazote foam, the other end of which is pinned with a larger headed pin. Micropins are headless and very fine and sharp, so need to be handled with forceps. When piercing the elytron take care that the pin does not spring out of the forceps as you may not be able to find it except by accidentally leaning your arm or treading on it!

Once mounted, labels need to be attached and whether hand-written or printed, should include collection date, name of location, grid reference (to 6 figures if possible), vice-county number, name of collector, brief mention of the habitat type and host plant (if known). If collected outside the UK, especially where mapping coverage is less thorough than the Ordnance Survey, more location detail may be required, often requiring latitude and longitude (e.g. from a GPS) or distance and bearing from a fixed point. Once identified, the species name (including author name e.g. '*Donacia simplex* Fabricius'), determiner's name and determination date should be added on a separate label, plus the sex if known. The importance of thorough and accurate labelling cannot be overstated. Without it, the specimen is nothing more than a dead beetle on a pin; with it, it is a valuable source of information that can be used for scientific, conservation and educational purposes.

Specimens then need to be stored in cartons, boxes or cabinets which are now generally lined with Plastazote, and it is worth acquiring the best quality you can afford, including looking for used storage, which may be of sounder construction than newer models. If lining your own containers, LD45 density Plastazote 10 mm thick is recommended. If you have limited resources or storage space, you can also donate specimens to museum collections, possibly retaining a small number of key specimens that you expect to use more regularly. In general, collections are organised in taxonomic order, following the most recent checklist, currently Duff (2012), though of course taxonomic changes do occur.

However you store your collection, you will need to ensure they are not damaged by pests, especially dermestid beetles of the genus *Anthrenus*, which feed on dry material

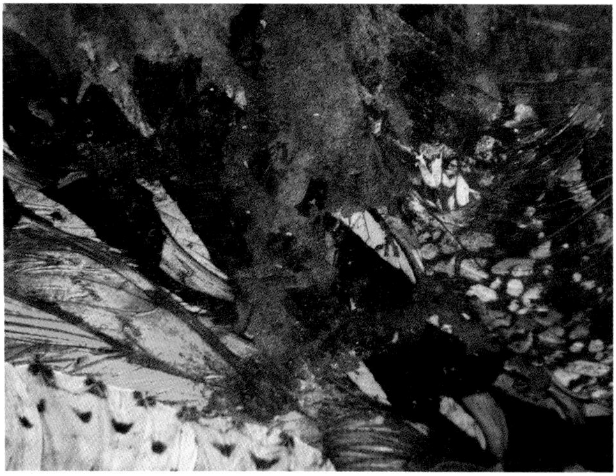

Fig. 7.9 Damage to butterfly specimens caused by dermestid beetles.

such as insect specimens (Fig. 7.9). Vigilance is essential, looking out for frass (insect faeces) in the storage container as well as the pests themselves, including shed skins. Many chemical deterrents such as naphthalene are toxic and thus banned outside the home; even though you may be allowed to use them privately, safer alternatives are preferred, in particular camphor, citronella, cedar or sandalwood oil. These have natural insecticidal properties and are placed in small, pinned, glass, fumigant cups, cleaning and refilling as necessary, with a small amount wiped around the rim of containers.

Storage boxes can be further protected by keeping them in sealed polythene bags, while cabinets should have seals around the doors and these should be checked periodically. If you notice or suspect pests in your collection, the best method of control is to wrap the affected box or drawer in polythene and put it in a freezer for two to three weeks.

So far, I have only referred to curation of adult specimens as most entomologists only retain collections of these. However, if you study juvenile stages, as I do, they need to be preserved in 70–80% alcohol (again, see Chapter 8 for details of equipment suppliers) in small, sealed tubes. Labels should be placed inside the tube as well which means the ink needs to be insoluble in alcohol – if you are unsure, make some trial labels and avoid the inks that run or dissolve. To avoid evaporation, even from sealed tubes, they can be stored upside down in alcohol in larger sealed containers. This method can also be used to store unmounted and undetermined adult specimens for long periods without risking pest damage.

7.2 Raising and culturing

I do not intend to cover breeding and culturing in any detail, although they may be undertaken occasionally, for example if a rare species is being bred in captivity for release to a new site, or to replenish a population that has declined locally. However, as with many insects, it can be difficult to identify juvenile stages, so if eggs, larvae or pupae are found, it may be necessary to raise them to adulthood for identification.

They can be kept in any suitable container. Plastic containers that are not air-tight, or have small air-holes, are fine, including Petri dishes and takeaway food containers. Eggs should be left on the section of substrate (generally part of the host plant) where they were found and will hatch after several days or a couple of weeks at most. After that time it is unlikely that any remaining eggs are viable and they may begin to darken and shrivel.

Larvae need fresh supplies of their foodplant, which can be put in a small container of water just as cut flowers are put in vases. Once pupated, the adult food supply (which may not always be the same as the larval one) needs to be introduced ready for emergence which may only take a few days. This can be the most difficult part of the process as you need to choose the correct plant. If you know which species it is, you can look up the food plant e.g. in Cox (2007), Hubble (2012) and the chrysomelid pages of the coleoptera.org.uk website. If you do not know which species it is, you will need to use the plant it was found on. Therefore it is a good idea to take a piece of the host plant when collecting larvae as they are not easy to identify. It is worth using tissue paper or similar to seal any gaps around the rim of the container or between stems to make sure beetles do not fall into the water and drown.

Once adults have emerged, they can hopefully be identified. If dissection is not required, live specimens can be released where the juveniles were found, unless it is an invasive species.

7.3 Dissecting

Unlike some groups of organisms such as butterflies and birds, leaf beetles can generally not be identified at a distance, and in many cases even a specimen in the hand will not be sufficient. Instead examination under a low-power microscope is often required, in many cases including dissection of the genitalia, especially in the Alticini. Most chrysomelids are small and the keys in Chapter 6 show that tiny features may need to be investigated closely. If the specimen is carded, the glue will need to be dissolved by floating the card in water to free the beetle for dissection. It can be re-glued afterwards and the genitalia added to the card behind the tip of the abdomen.

Fresh or relaxed specimens are needed for dissection. The last two abdominal segments can be removed with fine pins or forceps, or the abdomen squeezed to extrude the genitalia which can then be teased out. In males, it is usually the median lobe of the aedeagus that is needed for identification. As a hard, sclerotised structure, it can be teased away from the surrounding tissues and investigated with a binocular microscope before being mounted with the adult, along with any tergites and sternites that were removed. Identification keys generally provide line drawings of the median lobe where required to separate males of different species, but females may not be so readily separated by dissection. Fewer keys include details of female genitalia as the differences between species are less clear than in males, and the structures required (typically the spermatheca) less easy to dissect. However, some more specialist works, such as Konstantinov (1998), do make extensive use of the spermatheca to separate females and it is worth becoming familiar with this structure. Some guides to dissection procedure suggest using a solution of potassium hydroxide to remove unwanted soft tissue, but I have found the use of very fine mounted needles and warm water in a suitable dish to be sufficient.

tergite
upperside abdominal plate (sclerite) of a segment

sternite
underside abdominal plate (sclerite) of a segment

In some works, genetic information such as the diploid (2n) chromosome number may be given and there are undoubtedly chromosomal differences that could be investigated to help determine the taxonomic status of certain species and species complexes such as within the genus *Oulema* where the identity of several taxa is unclear. Assuming suitable material for comparison is available, the procedure for making chromosome preparations is not unduly difficult, but does require some laboratory equipment beyond that available to most coleopterists, including magnification at ×3,000. However, Cooter & Barclay (2006) again provide details should you wish to try this.

7.4 Studying

So far this chapter and the preceding keys have provided you with the information needed to collect and identify specimens. However, this is, in some ways, only the starting point. Once you have a specimen which has been confidently identified and includes all the associated information, you need to know what to do with it. In the UK, the first thing is to ensure your records are sent to the Leaf and Seed Beetle Recording Scheme (for Ireland, use www.biodiversityireland.ie). This is the national-scale scheme hosted by the Biological Records Centre (BRC) and includes a nationwide database of chrysomelid records going back to the 19th century. These were originally derived from specimen labels and hand-written lists, and then hard-copy recording cards, but now records are provided electronically as Excel spreadsheets or via direct online recording through systems such as iRecord. If using a spreadsheet, the following headings are used by the national scheme, and it is essential to include information for those in bold, and as many of the others as you can:

Species: The scientific name, not a common one, though few chrysomelids have common names in any case.

Locality: The name of the place where the specimen was found. Try to be specific e.g. 'North edge of Turnham Green' rather than 'Chiswick' or 'London'. If the name is given on an OS 1:50,000 map, then use this.

Vice-county (VC): These are county and sub-county divisions listed in many biological atlases and online e.g. South Hampshire and North Hampshire are separate VCs. They are useful as traditional ways of dividing areas for recording and many have their own recording scheme, especially for more popular species groups such as birds, plants and butterflies. If including this information, the VC number is all that is needed rather than the name.

Grid reference: The OS grid reference to 6 figures if possible (though 2- and 4-figure grid references can be accepted), but always including the two grid square letters e.g. 'TQ323587' not just '323587' as the

	number alone is not a unique location. If the specimen is associated with a very narrowly defined location, 8 figures may be appropriate if you can determine them using a sufficiently detailed map or a GPS.
Altitude:	In metres above sea level.
Day:	Day of the month as a number.
Month:	As a number from 1 to 12.
Year:	As a 4-digit number. (Usually a single date is given, and it is certainly preferred, but if giving a range, the Day, Month and Year relate to the starting day.)
Day2:	If you are providing a date range rather than a single day, this is the final day.
Month2:	The final month of a range.
Year2:	The final year of a range.
Recorder:	The name of the person who found the specimen.
Determiner:	The name of the person who identified the specimen. This is often the same as the recorder but not for example if a specialist has provided an identification.
Habitat:	Standard EUNIS (European Nature Information System) habitat codes can be used, otherwise a brief description such as 'deciduous woodland', 'urban garden', 'scrub on chalk grassland' or similar.
Habitat_2:	A second EUNIS habitat code (otherwise generally left blank).
Abundance:	Number present. An estimate may be given but this is generally left blank as the abundance is not known.
Comments:	Any other relevant information such as hostplant, sex, life history stage, more detailed information about the location and habitat, and anything else you feel would be useful.

The database can be accessed freely via the National Biodiversity Network (NBN) website, though some records are deemed sensitive. For example, a species may be potentially threatened by over-collection if it is highly localised, or in some cases there are confidentiality issues such as an invasive or pest species found at a commercial location.

Once records are accessed they can be used in many ways, for example by studying and mapping the distribution of species using the grid references included. One current work taking this approach is 'the Atlas' (Cox, 2007) which maps all the chrysomelid species in the British Isles.

Of course, some specimens will prove problematic and there are times when you need to consult an entomological collection or a specialist. In most cases, if you use a collection, it will be at a museum or an entomological organisation such as the British Entomological and Natural History Society (BENHS). Access varies, but in general you simply need to ask in advance and say why you want to use the collection. You will then be able to compare your specimen with some that have already been identified.

If looking for a specialist, you have several options. One is to take good-quality, high-resolution photographs and post them on a website such as iSpot where the recording community, from beginners to experts, can try to make an identification. You do need to know which features are important and photograph them; a simple dorsal view of the whole beetle is unlikely to be sufficient and you may need to include details of the head, tarsal segments, antennal segments or dissected genitalia. For example, *Longitarsus ballotae* is around 2 mm long but is separated from others in its genus by the presence of a small tooth at the base of each tarsal claw; a tiny feature indeed. Similarly, there are communities on social media such as Facebook's 'Coleoptera' group where specialists post images of beetles. This tends to be used by more-experienced, amateur, beetle workers.

However, in many cases, even clear magnified photographs are not sufficient and you may need a specialist to see your beetle specimen first-hand. The first place to ask is the organiser of the Recording Scheme whose details can be found on the BRC website. Remember that they run the scheme voluntarily and identify specimens in their 'spare' time, so do ask before sending specimens, and include sufficient stamps if you want your beetles returned to you. However, scheme organisers will generally help with identification when time allows.

7.5 Mark–release–recapture

If estimates of population size are required, a useful technique is mark–release–recapture (MRR). This involves capturing a number of beetles and marking the elytra with nail varnish, correcting fluid (e.g. Tippex) or a fast-drying cellulose-based paint. The marks are tiny dots applied with a cocktail stick, fine artist's brush, or similar fine point, avoiding the use of pins or other materials that might damage the beetle. Once the marks have dried and the beetles checked to ensure they have not stuck to anything or had appendages glued by the paint, they are released. In some cases, such as certain behavioural studies, beetles need to be individually recognisable. If so, dots are painted in combinations of different colours and positions on the elytra to give sufficient individual 'codes', much like the combinations of coloured leg-rings used to identify large wading birds at a distance. Similarly, different colours or patterns can be used to differentiate between different capture or release dates, or other variables as required.

After a period of time dependent on experimental design, a number of beetles are again captured, some of which will hopefully be marked specimens. The proportion of marked specimens recaptured then provides an estimate of population size:

Estimated total population (N) = (total number collected in second sample × number marked in first sample)/number of marked specimens recaptured.

As this is an estimate, it is important to consider the likely extent of variation from the true population size. For statistical purposes, 95% confidence limits are most commonly used:

$$N \pm 2N \sqrt{(1/R - 1/S_2)}$$

where:

N = estimated population size
R = number of marked specimens recaptured
S_2 = total number collected in second sample

This means that reliable estimates tend to require fairly large numbers of specimens. For example, if 50 *Gastrophysa viridula* were captured, marked and released, then 100 captured a week later, of which 10 were marked, the population estimate would be (100 × 50)/10 = 500. However, the 95% confidence limits would be 500 ± 1000√(1/10 – 1/100) = 500 ± 300, a range of 200–800. Thus, larger samples, of several hundred specimens, are often required if the variation is to be a smaller percentage of the total estimate.

This in turn means that the technique can be problematic if an estimate of a small population is required, as might be the case for scarce species, or those difficult to locate for other reasons. There are also a number of assumptions that need to be taken into account:

- The population does not change between samples i.e. no beetles enter, leave or die.
- All individual beetles in the population are equally likely to be captured.
- Marking does not affect assumption (2).
- A reasonable number of marked individuals are captured in the second sample.

To ensure these assumptions remain valid, the time between samples needs to be appropriate. It varies with time of year, weather and food supply, but as a rule-of-thumb, a period of around a week is sufficient. This allows marked individuals to mix with the general population but is unlikely to allow enough time for much migration, emergence or mortality.

However, carefully designed use of MRR techniques can provide information about aspects beyond simple population size, such as migration, adult emergence rate, mortality, longevity, male:female sex ratio, and individual behaviour. Further details of MRR techniques are given in many publications (e.g. Wheater & Cook, 2003) and online.

7.6 How to present your findings

If you study leaf beetles, you will soon encounter interesting specimens, unfamiliar behaviours and many other aspects that make such study worthwhile. These may not be new to science, but this does not mean you should not tell the wider world about them, maybe via a blog or local Natural History Society newsletter. There are many options. If you have records of the beetles you have identified, either add them via iRecord (www.brc.ac.uk/irecord), or contact the organiser of the Chrysomelid Recording Scheme to obtain a spreadsheet in the correct format for records to be added easily to the scheme database.

However, it is possible that you will find something worth publishing more formally, such as a species new to Britain or a part of Britain, a species on a new host, a new parasite, or previously undescribed behaviours or juvenile stages – and I hope you do! After all, this is how our knowledge of chrysomelids progresses. If so, you need to choose a suitable journal and, following their submission guidelines (which are available in a copy of the journal, online or by request), send them a short article. In Britain, *The Coleopterist* is suggested, but other options include the BENHS's *British Journal of Entomology and Natural History* and the Amateur Entomologists' Society's (AES) *Entomologist's Record and Journal of Variation*. See Chapter 8 for contact details of the relevant organisations. Those unfamiliar with publishing conventions are advised to examine current numbers of these journals to see what sort of thing they publish, and then to write a paper along similar lines. Keep it short, but present enough information to establish the conclusions. It is then time to consult an appropriate expert who can give advice on whether and in what form the material might be published. It is an unbreakable convention of scientific publication that results are reported with scrupulous honesty. Hence it is essential to keep detailed and accurate records throughout the investigation, and to distinguish in the write-up between certainty and probability, and between deduction and speculation. It will usually be necessary to apply appropriate statistical techniques to test the significance of the findings. A book such as *Studying Invertebrates* (Wheater & Cook, 2003) or the *Open University Project Guide* (Chalmers & Parker, 1989) will help, but this is an area where expert advice can contribute much to the planning, as well as the analysis, of the work.

7.7 Selected plants and their associated chrysomelids

A selection of plants and plant groups are given as a starting point for those who would like to use this to help focus their searching for chrysomelids. It is not, however, intended as a comprehensive list as almost any plant might yield these beetles, which as a family feed on a very wide range of plants. So, although they are not covered here in detail, gorses, St. John's-worts, buttercups, flax and linseed, rock-roses, purple-loosestrife, spurges, speedwells, willowherbs, bindweeds, comfreys and figworts are all well worth checking, as are many other plants. More complete coverage of foodplants is given in Cox (2007) and Hubble (2012). A number of chrysomelid species of particular interest are indicated that may inspire further study and which serve as examples of species feeding on the chosen plants.

Brassicas

This includes many wild plants as well as crops such

as cabbages, oilseed rape, mustards, turnips and watercresses. A number of chrysomelid species feed on plants in this family, especially the flea beetles, some of which are considered to be pests. For example, several *Phyllotreta* species such as *P. atra*, *P. chrysocephala*, *P. nemorum*, *P. nigripes* and *P. undulata* are well known on oilseed rape as well as being found on a range of other brassicas. If you search wild and cultivated members of the Brassicaceae, you are highly likely to find a range of flea beetles as well as others such as *Phaedon cochleariae*.

Grasses, including cereals

Like the brassica-feeders, several cereal-feeding species also take advantage of the intensive cultivation of their food source. *Oulema melanopus*, *Oulema obscura* and *Neocrepidodera ferruginea* in particular can be found on cereal crops and wild grasses. In Ireland, *Oulema septentrionis* can be found on oats but is not known from the rest of the British Isles. *Chaetocnema hortensis* is known from a wide range of wild and cultivated grasses.

Scrubby trees

Many chrysomelid species can be found on scrubby trees such as willows, hazels and birches among others, and using a stick and beating tray is likely to yield a range of specimens. It is a good way of finding some *Cryptocephalus* species, not only common ones such as *C. labiatus* and *C. pusillus*, but also, if you are fortunate, rarities like *C. coryli*, *C. decemmaculatus*, *C. frontalis* and *C. nitidulus*.

Others include *Plagiodera versicolora*, *Zeugophora*, *Chrysomela*, *Gonioctena*, *Phratora*, *Crepidodera* and *Luperus* species, *Altica brevicollis*, *Lochmaea caprea* and *Lochmaea crataegi*. The alder leaf beetle *Agelastica alni* is a species of particular interest that feeds on such vegetation. It was considered extinct by the late 20th century, but after being rediscovered breeding in the Manchester area in 2004, several further sightings were made, with a rapid expansion apparent in northern England during 2014. Also in 2014, a large number of individuals were recorded in southern Hampshire, indicating a rapid expansion and recolonisation in this area as well (Hubble, 2015). With records being sent regularly to the recording scheme from further afield at the time of writing, this process looks to be continuing. Rarer still, the only British specimen of *Smaragdina salicina* was found by beating this type of vegetation (Hubble & Murray,

2011), so it is certainly possible to make new discoveries as well as becoming more familiar with common species.

Plantains

A number of chrysomelids feed on *Plantago* species, including *Chrysolina haemoptera*, mainly on buck's-horn plantain around the coasts of southern Britain and the rare *Chrysolina latecincta* found on several plantains, but only in a few locations in Scotland. *Phaedon concinnus* and *Longitarsus plantagomaritimus* may be found on sea plantain around the coast, while *Apteropeda orbiculata*, *Longitarsus pratensis*, *Longitarsus kutscherae* and, less commonly *Longitarsus reichei*, are found on various plantains in many habitats.

Docks and sorrels

The most familiar chrysomelid associated with *Rumex* species is probably the green dock beetle *Gastrophysa viridula*, which is often seen as mating pairs or gravid females with swollen black abdomens, while the small orange eggs and black larvae are also commonly seen. *Mantura rustica* is scarce but widely scattered around the country, with *Mantura obtusata* and the rarer *Mantura chrysanthemi* also widely scattered, while *Chaetocnema concinna* is common and widespread.

The Pashford pot beetle *Cryptocephalus exiguus* is associated with common sorrel, but declined to a single site, Pashford Poors Fen in Suffolk where it has not been found since 2000 following drainage of surrounding agricultural land that dried the fenland habitat it relied on. It is therefore likely to be extinct in the UK, but is included here, not only as an example of how conservation needs to be at a landscape scale beyond simply small protected 'islands' of habitat, but also because it could be rediscovered if there are overlooked populations, or if it recolonised from continental Europe.

Asteraceae

This is the daisy family which is a large group of plants including ragworts, thistles, tansy, fleabanes, hawkweeds, knapweeds and yarrow, among many others. Thistles are important for *Lema cyanella*, *Sphaeroderma* species, *Galeruca tanaceti* and the rare *Galeruca laticollis*, with other Asteraceae important for *Longitarsus succineus*, *Longitarsus suturellus*, *Neocrepidodera transversa*, *Pilemostoma fastuosa* and several *Cassida* species. Some of the thistle-feeding species such as *Sphaeroderma* can also be found on other Asteraceae. The mainly yellow 'dandelion-like' hawkweeds and related species support *Cryptocephalus aureolus* and *Cryptocephalus hypochaeridis*.

Tansy is of particular importance for the rare tansy beetle *Chrysolina graminis*, although it is also known from water mint. Having declined from a scattered distribution around Britain to a stretch of the River Ouse in Yorkshire, its status was uncertain following floods along the Ouse in recent years which had the potential to wipe out its main British population. However, there has been a recent reintroduction to Wicken Fen in Cambridgeshire, and the beetle was also found in Woodwalton Fen, also Cambridgeshire, after an absence of some 40 years.

Lamiaceae

This family comprises the dead-nettles, mints and relatives and supports a range of chrysomelids including *Chrysolina fastuosa*, the rosemary beetle *Chrysolina americana*, and the mint leaf beetle *Chrysolina herbacea*. The rainbow leaf beetle *Chrysolina cerealis* feeds on wild thyme and in Britain is found at only a few protected reserve locations in Snowdonia. Others include many flea beetles of the genus *Longitarsus* such as the mint flea beetle *L. ferrugineus*, as well as *Apteropeda globosa* and the rare *Dibolia cynoglossi*, which has a stronghold in Dungeness and is sought on the coastal shingle.

Heathers and heaths

These plants do not support a large number of chrysomelids, but some do rely on such plants and would be under-recorded if they weren't checked for beetles. The most well known is probably the heather beetle itself, *Lochmaea suturalis*, which is sometimes considered a pest, especially on commercial grouse moors, while others include *Altica longicollis*.

Cryptocephalus biguttatus is a rarity associated with cross-leaved heath, and has only been recorded recently from a few heathland sites in southern England. It was previously known, albeit still rarely, from northern England and from a wider range of southern sites.

Water plants

These include members of several plant families such as sedges, reeds, bur-reeds, bulrushes, water-lilies, sweet-grasses and pondweeds. They are particularly important for the Donaciinae, most of which can be found on several plant species, although some are associated more closely with a single plant such as *Donacia clavipes* on common reeds, and *Donacia marginata* on branched bur-reed. Other beetles reliant on such plants are *Chaetocnema aerosa* and *Chaetocnema arida* which are found on spike-rushes, *Chaetocnema confusa* mainly on sedges, and *Chaetocnema sahlbergii* and *Chaetocnema subcoerulea* on sedges and rushes.

8 Useful addresses and links

8.1 Societies

Amateur Entomologists' Society (AES). PO Box 8774, London SW7 5ZG. Online contact via the form at www.amentsoc.org

One of the UK's leading organisations for people interested in insects. A charity run by volunteers that promotes the study of entomology, especially among amateurs and younger people, and produces a range of books, magazines and journals.

Buglife (The Invertebrate Conservation Trust). Bug House, Ham Lane, Orton Waterville, Peterborough PE2 5UU. Tel.: 01733 201210. E-mail: info@buglife.org.uk. www.buglife.org.uk

A British-based nature conservation charity focusing on invertebrates, including a range of campaigns and activities.

British Entomological and Natural History Society (BENHS). The Pelham-Clinton Building, Dinton Pastures Country Park, Davis Street, Hurst, Reading, Berkshire RG10 0TH. E-mail: enquiries@benhs.org.uk. www.benhs.org.uk

Promotes research in entomology, including conservation aspects. Runs activities such as lectures and field meetings and has a library and insect collections at its headquarters. Also publishes a quarterly journal as well as reference books.

Royal Entomological Society (RES). The Mansion House, Chiswell Green Lane, St Albans, Herts AL2 3NS. Tel.: 01727 899387. Online contact via the form at www.royensoc.co.uk.

An organisation that disseminates information about insects and improves communication between entomologists, both nationally and internationally. Publishes the Handbooks for the Identification of British Insects series. Out-of-print titles can be downloaded from the website. Includes a number of Special Interest Groups.

Watford Coleoptera Group (WCG). E-mail: thewcg@live.co.uk. www.thewcg.org.uk

A local association of coleopterists aiming to encourage enthusiasm for the study of beetles. The web site includes an excellent annotated gallery of many British species.

8.2 Suppliers of books and equipment

NHBS Ltd. 1-6 The Stables, Ford Road, Totnes TQ9 5LE. Tel.: 01803 865913. E-mail: customer.services@nhbs.com. www.nhbs.com

Books and equipment for wildlife, science and the environment.

Pemberley Books. 18 Bathurst Walk, Iver SL0 9AZ. Tel.: 01753 631114. Online contact via the form at www.pemberleybooks.com

Specialist natural history bookseller with a strong focus on entomology.

Watkins & Doncaster. PO Box 114, Leominster, Herefordshire HR6 6BS. Tel.: 0333 8003133 or 01568 750657. Online contact via the form at www.watdon.co.uk

Entomological equipment supplier.

8.3 Sources of data, journals and biological recording

Biological Records Centre (BRC). CEH Wallingford, Maclean Building, Crowmarsh Gifford, Wallingford, Oxfordshire OX10 8BB. Tel.: 01491 692564. E-mail: brc@ceh.ac.uk. www.brc.ac.uk

The national centre for species recording, working closely with the voluntary recording community, principally through support of national recording schemes and societies. Includes details of the Chrysomelid Recording Scheme and its organiser. Its work is a major part of the National Biodiversity Network.

National Biodiversity Network (NBN). Broadway Business Centre, 32a Stoney Street, Lace Market, Nottingham NG1 1LL. Tel.: 0115 9247132. E-mail: support@nbn.org.uk. www.nbn.org.uk

A central portal for sharing and accessing biological records, creating distribution maps, and uploading sightings to iRecord.

The Coleopterist. 8 Harvard Road, Ringmer, Lewes, East Sussex BN8 5HJ. www.coleopterist.org.uk

The leading journal for students of the beetle fauna of the British Isles.

GB Non-Native Species Secretariat (NNSS). Animal Health and Veterinary Laboratories Agency (AHVLA), Sand Hutton, York YO41 1LZ. E-mail: nnss@ahvla.gsi.gov.uk. www.nonnativespecies.org

A range of tools and information designed to help with the problem of invasive non-native species in Britain. Includes the searchable Non-Native Species Information Portal (NNSIP) with distribution and other data about non-native species, a gallery and identification material, action plans, risk analyses and species alerts that form part of the rapid response protocol. The NNSS is where to go if you need to report an invasive species that isn't already known about (for example if you discovered Colorado potato beetle living in Britain).

Beetles and beetle recording in Great Britain: Chrysomelidae. www.coleoptera.org.uk/family/chrysomelidae

The leaf beetle section of the wider British beetles website. Includes a photo gallery and information about individual British species. You can also upload sightings to iRecord.

Chrysomelidae: The Leaf Beetles of Europe and the Mediterranean Subregion. culex.biol.uni.wroc.pl/cassidae/European Chrysomelidae/index.htm.

A list of species in Europe where most will be illustrated by close-up photos of museum specimens, often including dissected-out genitalia and colour variants. Created by Dr Lech Borowiec, University of Wrocław, Poland. A well-respected online resource.

9 References and further reading

This book is intended as an introduction to the leaf beetles as a beginners' (or even improvers') guide to the biology of this group. There are more advanced texts such as Jolivet & Verma (2002), but these tend to be expensive and often difficult to find, though inter-library loans are a much cheaper way to access them, and your local library service can help with this. For readers wishing to widen their knowledge of the distribution and identification of British and Irish leaf beetles, Cox (2007) maps their distribution and Hubble (2012) gives keys to adults of all species. These books are readily available and can be seen as companion texts. To keep up to date with changes in our understanding of British beetles, including leaf beetles, *The Coleopterist* is an accessible and affordable journal forming a key part of the literature for serious study, whether amateur or professional.

Beyond this it becomes necessary to seek individual journal articles on particular species or genera of interest. There are also more technical books that cover wider areas such as Europe or the Palaearctic, and therefore include most or all of the British fauna within the wider range of species inhabiting the areas. For example, standard advanced texts, which can again be expensive and difficult to find, and are not all in English, include Mohr (1966) (central Europe, in German), Warchałowksi (2003) (Europe and the Mediterranean, in English) and Warchałowksi (2010) (the Palaearctic, in English). Doguet (1994) provides excellent coverage of French alticines and Winkelman (2014) covers Dutch chrysomelines, in French and Dutch respectively. Both include most British and Irish species.

These books all cover adult beetles. For juvenile stages, individual journal articles are again needed and Cox (2007) lists many of these, including several of his own key publications. There are no books providing keys to eggs or pupae, but Zaytsev & Medvedev (2009) provides keys (in Russian) to the larvae of Russian species, excluding the bruchines, and again covers many British and Irish species.

There are of course many online resources. A few are given in Chapter 8 as starting points, but web-searches may provide information that is not available in print. It is always worth checking, and the more you read, the more you will realise how many aspects of leaf beetles remain to be understood, and how many species there may be to discover, especially beyond the relatively well-researched fauna of Britain and continental Europe.

Balazuc, J. (1988) Laboulbeniales (Ascomycetes) parasitic on Chrysomelidae. In *Biology of Chrysomelidae*, eds. Jolivet, P., Petitpierre, E. & Hsiao, T.H., pp. 389–398. Dordrecht: Kluwer.

Baur, R. & Rank, N.E. (1996) Influence of host quality and natural enemies on the life history of the alder leaf beetles *Agelastica alni* and *Linaeidea aenea*. In *Chrysomelidae Biology. Vol. 2: Ecological Studies*, eds. Jolivet, P.H.A & Cox, M.L., pp. 173–194. Amsterdam: SPB Academic.

Breden, F. & Wade, M. (1985) The effect of group size and cannibalism rate on larval growth and survivorship in *Plagiodera versicolorea*. *Entomography* **3,** 455–463.

BRIG (2007) *Report on the Species and Habitat Review*. Report by the Biodiversity Reporting and Information Group (BRIG) to the UK Standing Committee, June 2007. Available from http://jncc.defra.gov.uk/PDF/UKBAP_Species-HabitatsReview-2007.pdf

Bukejs, A. (2014) A new species of the genus *Crepidodera* Chevrolat (Coleoptera: Chrysomelidae) from Baltic amber. *Zootaxa* **3815(2)**, 286–290.

Nora, C. Cabrera, N.C. & Durante, S.P. (2003) Comparative morphology of mouthparts in species of the genus *Acalymma* Barber (Coleoptera, Chrysomelidae, Galerucinae). *The Coleopterists Bulletin* **57(1)**, 5–16.

Castilla, A.M., Bauwens, D. & Llorente, G.A. (1991) Diet composition of the lizard *Lacerta lepida* in Central Spain. *Journal of Herpetology* **25(1)**, 30–36.

Chalmers, N. & Parker, P. (1989) *The OU Project Guide* (2nd ed.). Preston Montford: Field Studies Council.

Churchfield, S. & Rychlik, L. (2006) Diets and coexistence in *Neomys* and *Sorex* shrews in Białowieża forest, eastern Poland. *Journal of Zoology* **269**, 381–390.

Čižek, P. & Doguet, S. (2008) *Klíč k určování dřepčíků (Coleoptera: Chrysomelidae: Alticinae) Česka a Slovenska*. Nové Město nad Metují: Městské Muzeum.

Cooter, J. & Barclay, M.V.L. (eds.) (2006) *A Coleopterist's Handbook* (4th ed.). Orpington: Amateur Entomologists' Society.

Cox, M.L. (1981) Notes on the biology of *Orsodacne* Latreille with a subfamily key to the larvae of the British Chrysomelidae (Coleoptera). *Entomologist's Gazette* **32**, 123–135.

Cox, M.L. (1988) Egg bursters in the Chrysomelidae, with a review of their occurrence in the Chrysomeloidea and Curculionoidea (Coleoptera). *Systematic Entomology* **13,** 393–432.

Cox, M.L. (1996) Insect predators of Chrysomelidae. In *Chrysomelidae Biology. Vol. 2: Ecological Studies*, eds. Jolivet, P.H.A & Cox, M.L., pp. 23–92. Amsterdam: SPB Academic.

Cox, M.L. (2004) Flight in seed and leaf beetles (Coleoptera, Bruchidae, Chrysomelidae). In *New Developments in the Biology of Chrysomelidae*, eds. Jolivet, P., Santiago-Blay, J.A. & Schmitt, M., pp. 353–393. Amsterdam: SPB Academic.

Cox, M.L. (2007) *Atlas of the Seed and Leaf Beetles of Britain and Ireland*. Newbury: Pisces.
Cox, M. L. (2015) *Longitarsus fowleri* Allen, 1967 synonymised with *L. strigicollis* Wollaston, 1864 (Chrysomelidae). *The Coleopterist* **24(2)**, 93–99.
Daccordi, M. (1996) Notes on the distribution of the Chrysomelinae and their possible origin. In *Chrysomelidae Biology. Vol. 1: The Classification, Phylogeny and Genetics*, eds. Jolivet, P.H.A & Cox, M.L., pp. 399–412. Amsterdam: SPB Academic.
Degrugiller, M.E. (1996) Ultrastructure and distribution of *Rickettsia*-like organisms in reproductive tissues of the western corn rootworm, *Diabrotica virgifera virgifera*. In *Chrysomelidae Biology. Vol. 2: Ecological Studies*, eds. Jolivet, P.H.A & Cox, M.L., pp. 135–138. Amsterdam: SPB Academic.
Diaz, R., Hibbard, K., Samayoa, A. & Overholt, W.A. (2012) Arthropod community associated with tropical soda apple and natural enemies of *Gratiana boliviana* (Coleoptera: Chrysomelidae) in Florida. *Florida Entomologist* **95**, 228–232.
Doguet, S. (1994) Coléoptères Chrysomelidae. Volume 2. Alticinae. *Faune de France* **80**, 1–694.
Donaldson, J. R., & Lindroth, R. L. (2007) Genetics, environment, and their interaction determine efficacy of chemical defense in trembling aspen. *Ecology* **88**, 729–739.
Donisthorpe, H. St.J. K. (1902) The life history of *Clytra quadripunctata* L. *Transactions of the Entomological Society of London* **1902**, 11–24.
Duff, A.G. (ed.) (2012) *Checklist of Beetles of the British Isles* (2nd ed.). Iver: Pemberley Books.
Düngelhoef, S. & Schmitt, M. (2006) Functional morphology of copulation in Chrysomelidae-Criocerinae and Bruchidae (Insecta: Coleoptera). *Bonner Zoologische Beiträge* **54(4)**, 201–208.
Elias, S.A. & Kuzmina, S. (2008) Response of Chrysomelidae to Quaternary environmental changes. In *Research on Chrysomelidae Volume 1*, eds. Jolivet, P., Santiago-Blay, J.A. & Schmitt, M., pp. 174–193. Leiden: Brill.
Furth, D.G. (1988) The jumping apparatus of flea beetles (Alticinae) – The metafemoral spring. In *Biology of Chrysomelidae*, eds. Jolivet, P., Petitpierre, E. & Hsiao, T.H., pp. 285–298. Dordrecht: Kluwer.
Gathmann, A., & Tscharntke, T. (1999) Landschafts-Bewertung mit Bienen und Wespen in Nisthilfen: Artenspektrum, Interaktionen und Bestimmungsschlüssel. *Naturschutz und Landschaftspflege Baden-Württemberg* **73**, 277–305.
Ghent, A.W. (1960) A study of the group-feeding behaviour of larvae of the Jack pine sawfly, *Neodiprion pratti banksianae* Roh. *Behaviour* **16(1–2)**, 110–148.
Gibson, C.W.D. (1998) Brownfield: red data. The values artificial habitats have for uncommon invertebrates. *English Nature Research Reports* **273**, 1–38.
Glynn, C., Rönnberg-Wästljung, A.-C., Julkunen-Tiitto, R., & Weih, M. (2004) Willow genotype, but not drought treatment, affects foliar phenolic concentrations and leaf beetle resistance. *Entomologia Experimentalis et Applicata* **113**, 1–14.
Goidanich, A. (1956) Gregarismi od individualismi larvali e cure materne nei crisomelidi (Col. Chrysomelidae). *Memorie della Societa Entomologica Italiana* **35**, 151–182.
Grégoire, J.-C. (1988) Larval gregariousness in the

Chrysomelidae. In *Biology of Chrysomelidae*, eds. Jolivet, P., Petitpierre, E. & Hsiao, T.H., pp. 253–260. Dordrecht: Kluwer.

Gross, J., Fatouros, N. E., Neuvonen, S., & Hilker, M. (2004) The importance of specialist natural enemies for *Chrysomela lapponica* in pioneering a new host plant. *Ecological Entomology* **29**, 584–593.

Hubble, D. (2012) *Keys to the Adults of Seed and Leaf Beetles of Britain and Ireland*. Telford: FSC Publications.

Hubble, D.S. (2014) A review of the scarce and threatened beetles of Great Britain: The leaf beetles and their allies (NECR161). Natural England Commissioned Report. *Species Status* **19**, 1–118. Available from http://publications.naturalengland.org.uk/publication/6548461654638592

Hubble, D. (2015) The distribution of *Agelastica alni* (Linnaeus) (Chrysomelidae) in Hampshire and its relevance to the species' British status. *The Coleopterist* **24(1)**, 53–55.

Hubble, D. & Murray, D. (2011) First British record of *Smaragdina salicina* (Scopoli, 1763) (Chrysomelidae). *The Coleopterist* **20(1)**, 1–3.

Hubweber, L. & Schmitt, M. (2006) Parameres – similarities and differences in Chrysomelidae and Cerambycidae (Coleoptera). *Bonner Zoologische Beiträge* **54(4)**, 253–259.

Humber, R.A. (1996) Fungal pathogens of the Chrysomelidae and prospects for their use in biological control. In *Chrysomelidae Biology. Vol. 2: Ecological Studies*, eds. Jolivet, P.H.A. & Cox, M.L., pp. 93–116. Amsterdam: SPB Academic.

Hyman, P.S. & Parsons, M.S. (1992) *A Review of the Scarce and Threatened Coleoptera of Great Britain. Part 1*. JNCC: Peterborough.

Ikonen, A. (2002) Preferences of six leaf beetle species among qualitatively different leaf age classes of three Salicaceous host species. *Chemoecology* **12**, 23–28.

Invertebrate Link (2002) A Code of Conduct for collecting insects and other invertebrates. *British Journal of Entomology and Natural History* **15(1)**, 1–6. Available from http://www.royensoc.co.uk/InvLink/documents/Collecting Code (2002).pdf

JNCC & Defra (on behalf of the Four Countries' Biodiversity Group) (2012) *UK Post-2010 Biodiversity Framework, July 2012*. Available from http://jncc.defra.gov.uk/page-6189

Jolivet, P. & Verma, K.K. (2002) *Biology of Leaf Beetles*. Andover: Intercept.

Klein, M., Braverman, Y., Chizov-Ginzburg, A., Gol'berg, A., Blumberg, D., Khanbegyan, Y. & Hackett, K.J. (2002) Infectivity of beetle spiroplasmas for new host species. *BioControl* **47(4)**, 427–433.

Kölsch, G., Matz-Grund, C. & Pedersen, B.V. (2009) Ultrastructural and molecular characterization of endosymbionts of the reed beetle genus *Macroplea* (Chrysomelidae, Donaciinae), and proposal of "Candidatus Macropleicola appendiculatae" and "Candidatus Macropleicola muticae". *Canadian Journal of Microbiology* **55(11)**, 1250–1260.

Konstantinov, A. (1998) *Revision of the Palearctic species of* Aphthona *Chevrolat and cladistic classification of the Aphthonini (Coleoptera: Chrysomelidae: Alticinae)*. Gainesville: Associated Publishers.

Köpf, A., Rank, N. E., Roininen, H., Julkunen-Tiitto, R., Pasteels, J. M., & Tahvanainen, J. (1998) Phylogeny and the evolution of host plant use and sequestration in the willow leaf beetle genus *Phratora* (Coleoptera: Chrysomelidae). *Evolution* **52**, 517–528.

Martinková, Z. & Honěk, A. (2004) *Gastrophysa viridula* (Coleoptera: Chrysomelidae) and biocontrol of *Rumex* – a review. *Plant, Soil & Environment* **50(1)**, 1–9.

Menzies, I.S. & Cox, M. L. (1996) Notes on the natural history, distribution and identification of British reed beetles. *British Journal of Entomology and Natural History* **9**, 137–162.

Mikhailov, Y.E. (2008) Body colouration in the leaf beetle genera *Oreina* Chevrolat and *Crosita* Motschulsky and trends in its variation. In *Research on Chrysomelidae*, eds. Jolivet, P., Santiago-Blay, J.A. & Schmitt, M., pp. 129–148. Leiden: Brill.

Mohr, K.H. (1966) *Chrysomelidae*. In *Die Käfer Mitteleuropas*, eds. Freude, H., Harde. K.W. & Lohse, G.A., pp. 95–280. Krefeld: Goecke and Evers.

Morris, W., Grevstad, F. & Herzig, A. (1996) Mechanisms and ecological functions of spatial aggregation in chrysomelid beetles. In *Chrysomelidae Biology. Vol. 2: Ecological Studies*, eds. Jolivet, P.H.A. & Cox, M.L, pp. 303–322. Amsterdam: SPB Academic.

Müller, C. & Hilker, M. (2004) Ecological relevance of fecal matter in Chrysomelidae. In *New Developments in the Biology of Chrysomelidae*, eds. Jolivet, P., Santiago-Blay, J.A. & Schmitt, M., pp. 693-708. Amsterdam: SPB Academic.

Nahrung, H. & Marohasy, J. (1997) Maternal frass is necessary for embryonic development in *Weiseana barkeri* Jacoby (Coleoptera: Chrysomelidae). *Australian Journal of Entomology* **36(1)**, 95–96.

Nalepa, C.A. & Weir, A. (2007) Infection of *Harmonia axyridis* (Coleoptera: Coccinellidae) by *Hesperomyces virescens* (Ascomycetes: Laboulbeniales): role of mating status and aggregation behaviour. *Journal of Invertebrate Pathology* **94**, 196–203.

Nielsen, J.K. (1988) Crucifer-feeding Chrysomelidae: mechanisms of host-plant finding and acceptance. In *Biology of Chrysomelidae*, eds. Jolivet, P., Petitpierre, E. & Hsiao, T.H., pp. 25–40. Dordrecht: Kluwer.

Olmstead, K.L. (1996) Cassidine defenses and natural enemies. In *Chrysomelidae Biology. Vol. 2: Ecological Studies*, eds. Jolivet, P.H.A. & Cox, M.L, pp. 3–22. Amsterdam: SPB Academic.

Owen, J.A. (1999) Notes on the biology of *Cryptocephalus coryli* (Linnaeus) (Coleoptera: Chrysomelidae). *Entomologist's Gazette* **50**, 199–204.

Owen, J.A. (2003) Studies on the life-history of *Cryptocephalus nitidulus* Fabricius, 1787 (Coleoptera: Chrysomelidae). *Entomologist's Gazette* **54**, 255–266.

Palmén, E. (1944) Die anemohydrochore Ausbreitung der Insekten aus zoogeographischer Faktor, mit besonderer Berücksichtigung der baltischen Einwanderungsrichtung aus Ankunftsweg der fennoskandischen Käferfauna. *Annales Zoologici Societas Botanicae Fennicae, Vanamo, Helsinki* **10,** 1–262.

Palokangas, P. & Neuvonen, S. (1992) Differences between species and instars of leaf beetles in the probability to be preyed on. *Annales Zoologici Fennici* **29**, 273–278.

Pasteels, J.M., Braekman, J.-C. & Daloze, D. (1988) Chemical defense in the Chrysomelidae. In *Biology of Chrysomelidae*, eds.

Jolivet, P., Petitpierre, E. & Hsiao, T.H., pp. 233–252. Dordrecht: Kluwer.

Pasteels, J.M., Daloze, D. & Rowell-Rahier, M. (1986) Chemical defence in chrysomelid eggs and neonate larvae. *Physiological Entomology* **11**, 29–37.

Pasteels, J. M., & Gregoire, J.-C. (1984) Selective predation on chemically defended chrysomelid larvae. A conditioning process. *Journal of Chemical Ecology* **12**, 1693–1700.

Poinar, G.O. (1988) Nematode parasites of Chrysomelidae. In *Biology of Chrysomelidae,* eds. Jolivet, P., Petitpierre, E. & Hsiao, T.H., pp. 433–448. Dordrecht: Kluwer.

Poinar, G. & Jolivet, P. (2004) Origin of *Timarcha*: Trophic relationships in the Old and New World. In *New Developments in the Biology of Chrysomelidae,* eds. Jolivet, P., Santiago-Blay, J.A. & Schmitt, M., pp. 281–290. Amsterdam: SPB Academic.

Reid, C.A.M. (1995) A cladistic analysis of subfamilial relationships in the Chrysomelidae *sensu lato* (Chrysomelidae). In *Biology, Phylogeny, and Classification of Coleoptera: Papers Celebrating the 80th Birthday of Roy A. Crowson,* eds. Pakaluk, J. & Slipinski, S.A., pp. 559–631. Warszawa: Muzeum I Instytut Zoologii PAN.

Reid, C.A.M. & de Little, D.W. (2013) A new species of *Paropsisterna* Motschulsky, 1860, a significant pest of plantation eucalypts in Tasmania and Ireland (Coleoptera: Chrysomelidae: Chrysomelinae). *Zootaxa* **3681(4),** 395–404.

Roy, H.E., Brown, P.M.J., Comont, R.F., Poland, R.L. & Sloggett, J.J. (2013) *Ladybirds. Naturalists' Handbooks 10.* (2nd ed). Exeter: Pelagic Publishing.

Schaffner, U. & Müller, C. (2001) Exploitation of the fecal shield of the lily leaf beetle, *Lilioceris lilii* (Coleoptera: Chrysomelidae), by the specialist parasitoid *Lemophagus pulcher* (Hymenoptera: Ichneumonidae). *Journal of Insect Behavior* **14(6)**, 739–757.

Schmitt, M. (1988) The Criocerinae: biology, phenology and evolution. In *Biology of Chrysomelidae,* eds. Jolivet, P., Petitpierre, E. & Hsiao, T.H., pp. 475–495. Dordrecht: Kluwer.

Schmitt, M. (1994) Stridulation in leaf beetles (Coleoptera, Chrysomelidae). In *Novel Aspects of the Biology of Chrysomelidae,* eds. Jolivet, P.H., Cox, M.L. & Petitpierre, E., pp. 319–325. Dordrecht: Kluwer.

Schmitt, M. (2004) Jumping flea beetles: structure and performance (Insecta, Chrysomelidae, Alticinae). In *New Developments in the Biology of Chrysomelidae,* eds. Jolivet, P., Santiago-Blay, J.A. & Schmitt, M., pp. 161–170. Amsterdam: SPB Academic.

Selman, B.J. (1988) Chrysomelids and ants. In *Biology of Chrysomelidae,* eds. Jolivet, P., Petitpierre, E. & Hsiao, T.H., pp. 463–474. Dordrecht: Kluwer.

Shirt, D.B. (ed.) (1987) *British Red Data Books: 2. Insects.* NCC: Peterborough.

Stojanova, A. M. & Mollov, I. A. (2008) Diet and trophic niche overlap of the moor frog (*Rana arvalis* Nilsson, 1842) and the common frog (*Rana temporaria* L., 1758) from Poland. In *Proceedings of the anniversary scientific conference of ecology, Plovdiv,* eds. Velcheva, I.G. & Tsekov, A.G., pp. 181–190.

Tallamy, D. W., Hibbard, B. E., Clark, T. L. & Gillespie, J. J. (2005) Western corn rootworm, cucurbits, and cucurbitacins. In *Ecology and management of western corn rootworm* (Diabrotica

virgifera virgifera *LeConte*), eds. Vidal, S., Kuhlmann, U. & Edwards, R., pp. 67–93. Wallingford: CAB International.
Théodoridès, J. (1988) Gregarines of Chrysomelidae. In *Biology of Chrysomelidae*, eds. Jolivet, P., Petitpierre, E. & Hsiao, T.H., pp. 417–432. Dordrecht: Kluwer.
Toguebaye, B.S., Marchard, B. & Bouix, G. (1988) Microsporidia of the Chrysomelidae. In *Biology of Chrysomelidae*, eds. Jolivet, P., Petitpierre, E. & Hsiao, T.H., pp. 399–416. Dordrecht: Kluwer.
Tosun, O., Yaman, M. & Aydin, Ç. (2008) Parasites of *Phyllotreta atra* (Fabricius, 1775) (Coleoptera: Chrysomelidae) in Trabzon. *Türkiye Parazitoloji Dergisi* **32(2)**, 153–157.
Tsubaki, Y. (1981) Some beneficial effects of aggregation in young larvae of *Pryeria sinica* Moore (Lepidoptera: Zygaenidae). *Researches on Population Ecology* **23**, 156–167.
Unwin, D.M. (1984) *A key to the families of British beetles*. Preston Montford: Field Studies Council. Reprinted (1988) with minor amendments from *Field Studies* **6(1)**, 149-197.
Vencl, F.V. & Nishida. K. (2008) A new gall-inducing shining leaf beetle (Coleoptera: Chrysomelidae) from Thailand and its relevance to the evolution of herbivory in leaf beetles. In *Research on Chrysomelidae*, eds. Jolivet, P., Santiago-Blay, J.A. & Schmitt, M., pp. 246–259. Leiden: Brill.
Wade, M. J. & Breden, F. (1986) Life history of natural populations of the imported willow leaf beetle, *Plagiodera versicolora* (Coleoptera: Chrysomelidae). *Annals of the Entomological Society of America* **79**, 73–79.
Wallace, J. B. & Blum, M. S. (1969) Refined defensive mechanisms in *Chrysomela scripta*. *Annals of the Entomological Society of America* **62**, 503–506.
Wallace, J. B. & Blum, M.S. (1971) Reflex bleeding: A highly refined defensive mechanism in *Diabrotica* larvae (Coleoptera: Chrysomelidae). *Annals of the Entomological Society of America* **64**, 1021–1024.
Warchałowksi, A. (2003) *The Leaf-beetles (Chrysomelidae) of Europe and the Mediterranean Area*. Warsaw: Natura Optima Dux Foundation.
Warchałowksi, A. (2010) *Palaearctic Chrysomelidae. Identification Keys* (2 vols.). Warsaw: Natura Optima Dux Foundation.
Weselok, R. M. (1981) Host location by parasitoids. In *Semiochemicals. Their role in pest control*, eds. Norlund, D.A., Jones, R.L. & Lewis, W.J., pp. 79–95. New York: Wiley.
Wheater, C.P. & Cook, P.A. (2003) *Studying Invertebrates. Naturalists' Handbooks 28*. Slough: The Richmond Publishing Co. Ltd.
Wickler, W. (1968) *Mimicry in Plants and Animals*. New York: McGraw-Hill.
Winkelman, J. (2014) *De Nederlandse goudhaantjes (Chrysomelidae: Chrysomelinae)*. Leiden: EIS.
Zaytsev, Y.M. & Medvedev, L.N. (2009) *Lichinki zhukov-listoedov Rossii*. Moscow: KMK Scientific.
Zvereva, E. L., & Rank, N. E. (2004) Fly parasitoid *Megaselia opacicornis* uses defensive secretions of the leaf beetle *Chrysomela lapponica* to locate its host. *Oecologia* **140**, 516–522.

Index

Page numbers in *italics* denote figures and in **bold** denote tables.